# Green Energy Solutions

*A Guide to Practical Strategies for Implementing Renewable Energy Technologies in Your Home and Energy Storage*

## Lewis Finan

# Table of Contents

Introduction .................................................................................................8

Chapter 1: Introduction to Green Energy ...................................................11

  1.1 The Importance of Renewable Energy...................................................13

  1.2 Overview of Green Energy Technologies.............................................15

  1.3 Benefits of Implementing Green Energy at Home............................17

  1.4 Understanding Energy Storage .............................................................20

Chapter 2: Solar Energy Solutions .............................................................23

  2.1 Basics of Solar Power .............................................................................25

  2.2 Types of Solar Panels and Their Efficiency ......................................27

  2.3 Installing Solar Panels: Step-by-Step Guide ....................................30

  2.4 Cost and Maintenance Considerations .............................................34

Chapter 3: Wind Energy for Residential Use ............................................38

  3.1 Introduction to Wind Energy ..............................................................41

  3.2 Types of Residential Wind Turbines ..................................................44

  3.3 Site Assessment and Installation Process .........................................47

  3.4 Advantages and Limitations of Wind Energy ..................................51

Chapter 4: Geothermal Energy Systems ...................................................55

  4.1 Understanding Geothermal Energy ....................................................59

  4.2 Types of Geothermal Systems .............................................................63

  4.3 Installation and Maintenance of Geothermal Heat Pumps ..........67

  4.4 Environmental and Economic Benefits .............................................70

Chapter 5: Biomass and Bioenergy ...........................................................74

  5.1 Basics of Biomass Energy .....................................................................77

  5.2 Types of Biomass and Their Applications .........................................81

  5.3 Home Biomass Systems: Setup and Management ..........................86

  5.4 Sustainability and Environmental Impact ........................................90

Chapter 6: Energy Storage Solutions .........................................................94

  6.1 Importance of Energy Storage .............................................................98

  6.2 Types of Energy Storage Systems (Batteries, Thermal, Mechanical)...........101

  6.3 Selecting the Right Storage System for Your Home .......................105

  6.4 Integration with Renewable Energy Sources...................................108

Chapter 7: Smart Home Technologies and Energy Management.................................................113

    7.1 Overview of Smart Home Energy Systems ........................................................116

    7.2 Smart Grids and Their Role in Energy Efficiency...........................................121

    7.3 Home Automation for Energy Conservation ..................................................125

    7.4 Monitoring and Managing Energy Use ...........................................................129

Chapter 8: Financing and Incentives for Green Energy Projects .........................................135

    8.1 Understanding the Costs and Savings...............................................................139

    8.2 Government Incentives and Rebates.................................................................143

    8.3 Financing Options and Strategies .....................................................................147

    8.4 Long-term Financial Benefits of Green Energy Investments ........................152

Conclusion: ...................................................................................................................157

# Introduction

In an age where climate change and environmental degradation dominate global conversations, the quest for sustainable living has never been more urgent. As our planet faces unprecedented challenges, individuals, communities, and nations increasingly seek ways to reduce their carbon footprints and transition to cleaner, greener energy sources. "Green Energy Solutions: A Guide to Practical Strategies for Implementing Renewable Energy Technologies in Your Home and Energy Storage" aims to be a comprehensive resource for anyone ready to embark on this crucial journey.

This book is designed for homeowners, DIY enthusiasts, and anyone passionate about making a tangible difference in the fight against climate change. Whether you're a seasoned environmentalist or new to renewable energy, you'll find valuable insights and practical advice to guide you through transforming your home into an energy-efficient, eco-friendly haven.

In "Green Energy Solutions," we explore a variety of renewable energy technologies, from solar panels and wind turbines to geothermal systems and micro-hydro power. Each chapter delves into the specifics of how these technologies work, the benefits they offer, and the steps required to implement them in your home. We also provide detailed guidance on energy storage solutions, ensuring you can maximize the efficiency and reliability of your renewable energy systems.

This guide is not just about the technical aspects of green energy; it also addresses the broader context of sustainable living. We discuss the financial incentives and government programs available to support your transition, the environmental and economic benefits of renewable energy, and how adopting these technologies can enhance your quality of life.

By the end of this book, you will have a clear understanding of the various renewable energy options available, the knowledge to assess which solutions are best suited to your needs, and the confidence to take actionable steps towards a greener future. The journey to sustainable living starts with informed decisions and practical actions, and "Green Energy Solutions" is here to empower you every step of the way.

Join us as we explore the exciting world of renewable energy and discover how you can contribute to a healthier planet, one home at a time. Together, we can make a difference—one that ensures a brighter, cleaner future for generations to come.

# Chapter 1: Introduction to Green Energy

The modern world is at a critical juncture, grappling with the severe impacts of climate change and environmental degradation. As the energy demand continues to grow, the traditional reliance on fossil fuels has become increasingly unsustainable, contributing to pollution, greenhouse gas emissions, and resource depletion. In this context, the transition to green energy—renewable, clean, and sustainable energy sources—has emerged as a vital solution for securing a healthy future for our planet.

Green energy encompasses a variety of technologies that harness natural processes to generate power without depleting resources or causing significant environmental harm. Unlike fossil fuels, which release large amounts of carbon dioxide and other pollutants when burned, renewable energy sources like solar, wind, geothermal, and hydropower produce minimal or no emissions. This makes them crucial allies in the fight against climate change, as they help reduce the global carbon footprint and mitigate the adverse effects of industrial activities on the environment.

Solar energy, one of the most well-known forms of green energy, captures sunlight using photovoltaic cells or solar thermal systems. These technologies convert sunlight into electricity or heat, providing a versatile and abundant source of power. Wind energy, harnessed through turbines, converts the kinetic energy of wind into electricity. Wind farms, both onshore and offshore, have become increasingly common as advancements in technology have made wind power more efficient and cost-effective.

Geothermal energy taps into the Earth's internal heat, using it to generate electricity or provide direct heating. This energy source is particularly valuable because it is available around the clock, unlike solar and wind energy, which are dependent on weather conditions. Hydropower, the oldest form of renewable energy, generates electricity by capturing the energy of flowing water. Large-scale hydropower projects, such as dams, and small-scale installations like micro-hydro systems, offer reliable and sustainable energy solutions.

The adoption of green energy technologies is not only driven by environmental considerations but also by economic and social benefits. Renewable energy sources can reduce dependency on imported fuels, enhance energy security, and create jobs in emerging green industries. Moreover, the decentralized nature of many renewable energy systems, such as rooftop solar panels, empowers individuals and communities to generate their power, promoting energy independence and resilience.

Transitioning to green energy also necessitates a focus on energy efficiency and storage solutions. Energy efficiency involves using less energy to perform the same tasks, thereby reducing overall consumption and minimizing waste. This can be achieved through various means, including better insulation, energy-efficient appliances, and smart home technologies. Energy storage, on the other hand, addresses the intermittent nature of some renewable energy sources. Batteries and other storage technologies store excess energy produced during periods of high generation for use when production is low, ensuring a consistent and reliable energy supply.

Despite the clear advantages of green energy, the transition is not without challenges. Initial costs, regulatory hurdles, and the need for technological advancements are significant barriers that must be addressed. However, with growing awareness, supportive policies, and continued innovation, these obstacles are increasingly surmountable.

In conclusion, green energy represents a crucial pathway towards a sustainable and resilient future. By embracing renewable energy technologies and prioritizing energy efficiency and storage, we can significantly reduce our environmental impact, enhance energy security, and drive economic growth. This chapter serves as an introduction to the diverse and dynamic world of green energy, setting the stage for a deeper exploration of practical strategies for implementing these technologies in your home and beyond. The journey towards a greener future begins with understanding the potential of renewable energy and committing to making informed, sustainable choices.

## 1.1 The Importance of Renewable Energy

The importance of renewable energy in today's world cannot be overstated. As we confront the challenges of climate change, environmental degradation, and finite fossil fuel resources, renewable energy emerges as a beacon of hope for a sustainable future. The shift towards renewable energy sources is essential for multiple reasons, encompassing environmental, economic, and social dimensions.

First and foremost, renewable energy plays a pivotal role in combating climate change. The burning of fossil fuels for energy is the primary source of greenhouse gas emissions, which trap heat in the atmosphere and drive global warming. This warming has led to severe weather events, rising sea levels, and disruptions to ecosystems. Renewable energy sources, such as solar, wind, and hydropower, produce little to no greenhouse gas emissions. By replacing fossil fuels with these clean energy sources, we can significantly reduce the amount of carbon dioxide and other harmful pollutants released into the atmosphere, mitigating the impacts of climate change.

In addition to its environmental benefits, renewable energy also offers substantial economic advantages. The renewable energy sector has become one of the fastest-growing industries globally, creating millions of jobs in manufacturing, installation, and maintenance. These jobs often provide higher wages and better working conditions compared to those in the fossil fuel industry. Furthermore, renewable energy can reduce energy costs in the long run. While the initial investment in renewable technologies can be high, the operational costs are typically much lower than those of conventional power plants. Solar panels and wind turbines, for example, have no fuel costs and minimal maintenance expenses. This economic shift not only fosters job growth but also promotes energy affordability and stability.

Energy security is another crucial aspect underscoring the importance of renewable energy. Fossil fuel resources are concentrated in specific regions, often leading to geopolitical tensions and dependency on imports. Renewable energy, however, is abundant and can be harnessed locally, reducing reliance on foreign energy sources and enhancing national security. Countries with abundant renewable resources can achieve greater energy independence, insulating themselves from volatile global energy markets and supply disruptions.

Socially, the transition to renewable energy has the potential to improve public health and quality of life. Fossil fuel combustion releases pollutants such as sulfur dioxide, nitrogen oxides, and particulate matter, which contribute to respiratory and cardiovascular diseases. By shifting to cleaner energy sources, we can reduce air pollution and associated health problems, leading to fewer medical expenses and healthier communities. Additionally, the decentralization of energy production through small-scale renewable installations can empower communities, giving them control over their energy supply and fostering resilience against power outages and natural disasters.

Moreover, the environmental stewardship fostered by renewable energy can protect natural habitats and biodiversity. Renewable energy projects, when appropriately planned and implemented, have a lower ecological footprint compared to fossil fuel extraction and burning. Preserving ecosystems is vital for maintaining the balance of our natural world and ensuring the survival of countless species.

In conclusion, the importance of renewable energy is multifaceted, encompassing critical environmental, economic, and social benefits. By transitioning to renewable energy sources, we can combat climate change, stimulate economic growth, enhance energy security, improve public health, and protect our natural environment. This transition is not merely a choice but a necessity for ensuring a sustainable and prosperous future for generations to come. The urgency of adopting renewable energy cannot be ignored, and it is incumbent upon individuals, businesses, and governments to embrace this transformative change.

## 1.2 Overview of Green Energy Technologies

Green energy technologies are revolutionizing the way we produce and consume power, offering sustainable alternatives to fossil fuels. These technologies harness natural processes to generate electricity and heat, reducing environmental impact and promoting energy independence. This overview explores the key green energy technologies: solar, wind, geothermal, hydropower, and biomass, highlighting their principles, applications, and benefits.

Solar energy is one of the most accessible and widely used forms of green energy. It captures sunlight through photovoltaic (PV) cells, which convert solar radiation directly into electricity. These cells are often installed on rooftops or in large solar farms. Solar thermal systems, another application, use mirrors or lenses to concentrate sunlight, and heat fluids to produce steam that drives turbines for electricity generation. Solar energy is abundant and can be harnessed virtually anywhere, making it an ideal solution for both urban and rural areas. It also reduces electricity bills and lowers greenhouse gas emissions.

Wind energy utilizes the kinetic energy of moving air to generate electricity. Wind turbines, typically installed in wind farms on land or offshore, consist of blades that rotate when wind passes through them, driving a generator that produces electricity. Wind energy is one of the most cost-effective renewable energy sources, with large-scale wind farms capable of generating significant amounts of power. It is particularly effective in regions with strong, consistent winds. Wind energy reduces reliance on fossil fuels, cuts greenhouse gas emissions, and provides local employment opportunities in manufacturing and maintenance.

Geothermal energy taps into the Earth's internal heat, a stable and constant energy source. This heat can be accessed through geothermal power plants, which use steam produced from hot water reservoirs deep underground to drive turbines and generate electricity. Alternatively, geothermal heat pumps can provide heating and cooling for buildings by exploiting the relatively stable temperatures just below the Earth's surface. Geothermal energy is reliable and can operate continuously, unlike solar and wind, which are weather-dependent. It also has a small land footprint and minimal environmental impact compared to fossil fuel extraction.

Hydropower generates electricity by harnessing the energy of flowing or falling water. This is typically done through dams built on large rivers, where the water released from the reservoir flows through turbines, generating electricity. Small-scale hydropower systems, or micro-hydro, can be used in smaller streams and rivers, providing a flexible and sustainable energy solution for remote areas. Hydropower is highly efficient and can provide a consistent and reliable power supply. However, it requires significant initial infrastructure investment and can impact local ecosystems if not managed properly.

Biomass energy involves using organic materials, such as agricultural waste, wood chips, and even algae, to produce electricity, heat, or biofuels. Biomass can be burned directly for heat or converted into biogas or liquid biofuels through various chemical processes. This technology helps manage waste and reduces methane emissions from landfills. Biomass energy can be carbon-neutral if managed sustainably, as the carbon dioxide released during combustion is offset by the carbon dioxide absorbed by plants during their growth. It provides a versatile and renewable energy source, particularly in areas with abundant agricultural or forestry residues.

Each of these green energy technologies offers unique advantages and can be tailored to specific environmental and economic contexts. By integrating a mix of these technologies, we can build a more resilient and sustainable energy system. Transitioning to green energy not only mitigates climate change but also promotes energy security, creates jobs, and supports economic development. As technology advances and costs continue to decline, the adoption of green energy technologies will become increasingly feasible and essential for a sustainable future. This overview sets the stage for a deeper exploration of practical strategies for implementing these technologies in homes and communities, driving the shift towards a greener, more sustainable world.

## 1.3 Benefits of Implementing Green Energy at Home

Implementing green energy technologies at home offers numerous benefits, spanning environmental, economic, and social dimensions. As homeowners increasingly look for ways to reduce their carbon footprint, enhance energy efficiency, and gain energy independence, the adoption of renewable energy solutions becomes a compelling choice. This section explores the key advantages of integrating green energy into residential settings.

One of the primary benefits of implementing green energy at home is the significant reduction in environmental impact. Traditional energy sources, such as coal, oil, and natural gas, contribute heavily to greenhouse gas emissions, air pollution, and resource depletion. By switching to renewable energy sources like solar, wind, or geothermal, homeowners can substantially lower their carbon footprint. Solar panels, for example, produce electricity without emitting carbon dioxide or other harmful pollutants. This shift not only helps combat climate change but also improves local air quality, benefiting public health.

Economic savings are another major advantage of adopting green energy technologies. While the initial investment in renewable energy systems can be substantial, the long-term savings often outweigh these costs. Solar panels, for instance, can significantly reduce or even eliminate monthly electricity bills. Additionally, many governments offer incentives such as tax credits, rebates, and grants to offset the installation costs of renewable energy systems. Over time, as energy prices from traditional sources continue to rise, the cost savings from renewable energy can increase, providing a stable and predictable energy expense for homeowners.

Energy independence is a crucial benefit of implementing green energy at home. By generating their power, homeowners are less reliant on external energy suppliers and the volatility of fossil fuel markets. This independence provides a measure of security against energy price fluctuations and potential supply disruptions. In areas prone to power outages or natural disasters, having a home equipped with renewable energy systems and energy storage solutions, like batteries, ensures a more reliable power supply.

Green energy technologies also enhance the value of a property. Homes equipped with renewable energy systems, such as solar panels or geothermal heating, are often seen as more desirable in the real estate market. Buyers are increasingly looking for energy-efficient and sustainable homes, recognizing the long-term savings and environmental benefits these features offer. As a result, homes with green energy installations can command higher resale values and attract more prospective buyers.

Implementing green energy at home fosters a greater sense of environmental stewardship and community responsibility. Homeowners who adopt renewable energy solutions often become advocates for sustainability, influencing friends, neighbors, and the broader community to consider similar actions. This ripple effect can lead to wider adoption of green energy technologies, amplifying the collective impact on reducing environmental degradation and promoting sustainable living practices.

Moreover, the transition to green energy can improve the overall quality of life. Renewable energy systems, such as solar panels, operate quietly and with minimal maintenance compared to traditional power generators, which can be noisy and require frequent servicing. Enhanced energy efficiency measures, such as improved insulation and energy-efficient appliances, contribute to a more comfortable and healthier living environment by maintaining consistent indoor temperatures and reducing exposure to harmful pollutants.

Lastly, adopting green energy at home aligns with broader societal goals of sustainable development and resilience. As cities and countries work towards reducing their carbon footprints and meeting international climate targets, individual actions at the household level play a crucial role. By contributing to these efforts, homeowners not only help achieve national and global environmental goals but also inspire others to participate in the transition towards a more sustainable future.

In conclusion, implementing green energy at home offers a multitude of benefits, including reduced environmental impact, economic savings, energy independence, increased property value, community influence, improved quality of life, and alignment with sustainability goals. As the urgency to address climate change grows, the adoption of renewable energy technologies in residential settings becomes not just a personal choice but a necessary step towards a more sustainable and resilient world.

## 1.4 Understanding Energy Storage

Energy storage plays a crucial role in the efficient use of renewable energy sources. As renewable energy generation, particularly from solar and wind, can be intermittent, having a reliable means to store excess energy ensures a consistent and dependable power supply. Understanding the various energy storage technologies and their applications is essential for maximizing the benefits of renewable energy at home.

At the core of energy storage is the ability to capture energy produced at one time for use at a later time. This capability is vital for balancing supply and demand, especially when renewable energy sources are not generating power. For instance, solar panels generate electricity during daylight hours, but energy consumption peaks in the evening. Without storage, the surplus energy generated during the day would be wasted. Energy storage systems, therefore, bridge this gap, ensuring that the energy produced is available when needed.

Batteries are the most common form of energy storage for residential use. Lithium-ion batteries, in particular, have gained popularity due to their high energy density, efficiency, and declining costs. These batteries store electrical energy as chemical energy and convert it back to electricity when needed. Home battery systems, such as the Tesla Powerwall, can be integrated with solar panels to store excess solar energy generated during the day and supply power during the night or cloudy days. This setup enhances energy self-sufficiency, reduces reliance on the grid, and can provide backup power during outages.

Lead-acid batteries are another type of storage, often used in off-grid applications. While they are less expensive than lithium-ion batteries, they have a shorter lifespan, lower energy density, and require more maintenance. However, they remain a viable option for homeowners looking for a cost-effective storage solution.

Flow batteries are an emerging technology that offers advantages in terms of scalability and longevity. These batteries use liquid electrolytes that flow through a cell stack to store energy. They are particularly well-suited for larger-scale storage needs, as their capacity can be easily increased by adding more electrolyte solutions. While currently more expensive than lithium-ion and lead-acid batteries, flow batteries have the potential for significant cost reductions and wider adoption in the future.

Beyond batteries, thermal energy storage is another important method. This technology involves storing excess energy as heat, which can later be converted back into electricity or used directly for heating purposes. One common approach is using molten salt in concentrated solar power (CSP) plants. The salt retains heat efficiently, which can be used to produce steam and generate electricity even when the sun is not shining. For residential applications, thermal storage can involve simpler systems, such as storing hot water in insulated tanks, which can be used for space heating or domestic hot water.

Pumped hydro storage is the largest and most established form of energy storage globally. This method involves using excess electricity to pump water from a lower reservoir to an upper reservoir. When electricity demand increases, the stored water is released back down, turning turbines to generate electricity. While highly effective for large-scale energy management, pumped hydro is less feasible for residential use due to its significant space and water requirements.

Compressed air energy storage (CAES) is another innovative technology, where excess electricity is used to compress air and store it in underground caverns or tanks. When energy is needed, the compressed air is released, expanded, and used to drive turbines to generate electricity. Though not commonly used in residential settings, CAES offers potential for future applications as technology advances.

Effective energy storage solutions are essential for maximizing the potential of renewable energy. By storing excess energy and making it available during periods of low generation, these systems ensure a reliable and stable energy supply. This not only enhances the efficiency and sustainability of renewable energy sources but also provides homeowners with greater energy independence and resilience. As technology continues to evolve, energy storage will play an increasingly vital role in the transition to a cleaner, more sustainable energy future.

# Chapter 2: Solar Energy Solutions

Solar energy solutions have become a cornerstone of the renewable energy revolution, offering a practical and effective way to harness the sun's power for electricity and heating. This chapter explores the various aspects of solar energy, including how solar technology works, its benefits, the different types of systems available, and practical steps for implementing solar energy solutions in your home.

At the heart of solar energy solutions are photovoltaic (PV) cells, which convert sunlight directly into electricity. These cells are made from semiconductor materials, typically silicon that generate an electric current when exposed to sunlight. PV cells are assembled into panels or modules, which can be installed on rooftops, ground-mounted, or integrated into building materials. The electricity generated by these panels can be used immediately, stored in batteries for later use, or fed back into the grid.

One of the primary benefits of solar energy is its abundance and sustainability. The sun provides more energy to the Earth in one hour than the entire world consumes in a year. By capturing even a fraction of this energy, we can significantly reduce our reliance on fossil fuels. Solar energy is also a clean power source, producing no greenhouse gas emissions during operation, which helps mitigate climate change and reduce air pollution.

There are two main types of solar energy systems for residential use: grid-tied and off-grid. Grid-tied systems are connected to the local utility grid, allowing homeowners to draw electricity from the grid when their solar panels are not producing enough power and to feed excess electricity back into the grid. This arrangement can lead to lower electricity bills and, in some cases, earn credits through net metering programs. Off-grid systems, on the other hand, are not connected to the utility grid and rely entirely on solar panels and battery storage to meet the home's energy needs. These systems are ideal for remote locations where grid access is unavailable or unreliable.

Another important aspect of solar energy solutions is solar thermal technology, which uses the sun's heat for domestic hot water, space heating, and even cooling. Solar thermal systems typically involve solar collectors that absorb sunlight and transfer the heat to a fluid, which is then circulated through a heat exchanger to provide hot water or heating. This technology is highly efficient and can significantly reduce energy costs associated with heating and cooling.

Implementing solar energy solutions in your home involves several key steps. First, assess your energy needs and determine the best type of solar system for your situation. Conduct an energy audit to identify ways to improve energy efficiency, as reducing your overall consumption can make your solar system more effective and economical. Next, evaluate your property's solar potential by considering factors such as roof orientation, shading, and available space. A professional solar installer can perform a site assessment and provide recommendations tailored to your home.

Once you have a clear understanding of your energy needs and solar potential, you can explore financing options. While the upfront cost of solar panels can be high, there are numerous incentives, rebates, and financing programs available to help offset these costs. Many homeowners opt for solar loans, leases, or power purchase agreements (PPAs), which can make solar energy more accessible without large initial investments.

After securing financing, choose a reputable solar installer to design and install your system. Look for installers with experience, certifications, and positive reviews. A quality installation ensures that your system operates efficiently and safely for years to come. Once your system is installed, you can start enjoying the benefits of solar energy, including lower utility bills, reduced carbon footprint, and increased energy independence.

In conclusion, solar energy solutions offer a powerful way to harness the sun's energy for residential use. By understanding the technology, benefits, and implementation process, homeowners can make informed decisions and take meaningful steps toward a sustainable energy future. Solar energy not only provides a reliable and clean power source but also contributes to a broader movement toward environmental stewardship and resilience.

## 2.1 Basics of Solar Power

Solar power harnesses the energy from the sun and converts it into electricity or heat, providing a clean and renewable source of energy for homes and businesses. Understanding the basics of solar power involves exploring how solar energy is captured, the technology behind solar panels, and the different types of solar power systems available.

At the core of solar power technology are photovoltaic (PV) cells, which are the fundamental building blocks of solar panels. PV cells are made from semiconductor materials, most commonly silicon. When sunlight strikes these cells, it excites electrons, creating an electric current. This process, known as the photovoltaic effect, is the basis for converting solar energy into electricity. PV cells are grouped in modules or panels to increase the amount of electricity generated.

Solar panels can be installed on rooftops, mounted on the ground, or integrated into building materials such as solar shingles. The electricity produced by solar panels is direct current (DC), which must be converted to alternating current (AC) to be compatible with the electricity grid and household appliances. This conversion is done by an inverter, a crucial component of any solar power system.

There are two main types of solar power systems: grid-tied and off-grid. Grid-tied systems are connected to the local utility grid, allowing for the exchange of electricity between the home and the grid. When solar panels generate more electricity than the home needs, the excess power is fed back into the grid, often earning the homeowner credits through net metering programs. Conversely, when the panels do not produce enough electricity, such as during nighttime or cloudy days, the home can draw power from the grid. This arrangement ensures a consistent power supply and can lead to significant savings on electricity bills.

Off-grid systems operate independently of the utility grid and rely entirely on solar panels and battery storage to meet the home's energy needs. These systems are particularly useful in remote locations where grid access is not available or is unreliable. Batteries store excess electricity generated during sunny periods, providing power when solar generation is low. Off-grid systems require careful planning to ensure that battery capacity and solar panel output are sufficient to meet the home's energy demands.

Solar thermal technology is another important aspect of solar power, focusing on harnessing the sun's heat rather than converting it to electricity. Solar thermal systems use solar collectors, typically installed on rooftops, to absorb sunlight and transfer the heat to a fluid, such as water or antifreeze. This heated fluid is then used for domestic hot water, space heating, or even cooling through solar absorption chillers. Solar thermal systems are highly efficient and can significantly reduce energy costs associated with heating and cooling.

Implementing solar power in a home involves several key steps. First, assessing the property's solar potential is crucial. This includes evaluating the roof's orientation, angle, and shading, as well as the available space for solar panel installation. A professional solar installer can conduct a site assessment and provide recommendations tailored to the specific property.

Next, understanding energy needs is essential. Conducting an energy audit helps identify ways to improve energy efficiency, which can reduce the size and cost of the solar power system required. It's also important to explore financing options. While the initial cost of solar panels can be substantial, numerous incentives, rebates, and financing programs are available to make solar power more affordable. Options such as solar loans, leases, and power purchase agreements (PPAs) can help manage upfront costs.

Choosing a reputable solar installer is a critical step. Look for installers with experience, certifications, and positive customer reviews. A quality installation ensures that the solar power system operates efficiently and safely for many years. Once the system is installed and connected, homeowners can begin to enjoy the benefits of solar power, including lower electricity bills, reduced carbon footprint, and increased energy independence.

In conclusion, the basics of solar power encompass understanding how PV cells convert sunlight into electricity, the differences between grid-tied and off-grid systems, and the role of solar thermal technology. With careful planning, assessment, and professional installation, homeowners can successfully implement solar power systems to achieve sustainable and cost-effective energy solutions.

## 2.2 Types of Solar Panels and Their Efficiency

Solar panels, also known as photovoltaic (PV) modules, are crucial components of solar power systems that convert sunlight into electricity. Understanding the different types of solar panels and their efficiency ratings is essential for selecting the most suitable option for residential or commercial installations.

## Monocrystalline Solar Panels

Monocrystalline solar panels are made from high-purity silicon crystals, typically sliced from cylindrical ingots of silicon. These panels have a uniform black appearance and are known for their efficiency and longevity. Monocrystalline cells have a higher efficiency rate compared to other types, typically ranging from 15% to 22%. This means they can convert a higher percentage of sunlight into electricity, making them ideal for installations where space is limited or efficiency is a priority.

## Polycrystalline Solar Panels

Polycrystalline solar panels are made from silicon crystals that are melted together to form wafers. They appear bluish due to the multiple crystals in the silicon structure. Polycrystalline panels are less expensive to manufacture than monocrystalline panels and have a lower efficiency rating, typically ranging from 13% to 16%. While they are slightly less efficient than monocrystalline panels, polycrystalline panels offer a cost-effective option for homeowners looking to install solar power systems without compromising on performance.

## Thin-Film Solar Panels

Thin-film solar panels are made from layers of photovoltaic materials applied onto a substrate, such as glass, plastic, or metal. These panels can be flexible and lightweight, offering versatility in installation options. Thin-film technology includes various materials such as amorphous silicon (a-Si), cadmium telluride (CdTe), and copper indium gallium selenide (CIGS). Each material has different efficiency levels, with amorphous silicon typically having lower efficiency (around 7% to 10%) compared to crystalline silicon panels, while CdTe and CIGS can achieve efficiencies of up to 22%. Thin-film panels are less efficient in converting sunlight into electricity but can perform better in low-light conditions and at high temperatures, making them suitable for certain applications where space or weight constraints are critical.

## Bifacial Solar Panels

Bifacial solar panels can capture sunlight from both the front and rear sides of the panel. These panels have transparent back sheets or are mounted on a dual-glass structure, allowing sunlight to pass through and be reflected onto the rear side of the cells. Bifacial panels can achieve higher efficiency gains compared to traditional monofacial panels, especially in environments with reflective surfaces, such as snow-covered ground or white rooftops. The efficiency of bifacial panels can vary widely depending on installation conditions and the reflectivity of the surrounding environment.

## Performance and Efficiency Considerations

When selecting solar panels, efficiency is a critical factor to consider. Higher efficiency panels produce more electricity per square meter of surface area, maximizing the power output from limited rooftop space. However, efficiency is not the only consideration. Factors such as cost, installation requirements, durability, and warranty terms also play important roles in choosing the right solar panels for your needs.

It's important to note that solar panel efficiency ratings are influenced by various factors, including temperature, shading, and installation angle. Manufacturers provide efficiency ratings based on standardized testing conditions, but real-world performance may vary. Working with a reputable solar installer can help ensure that panels are selected based on site-specific conditions and energy needs.

In conclusion, the choice of solar panels—whether monocrystalline, polycrystalline, thin-film, or bifacial—depends on factors such as efficiency goals, budget constraints, available space, and installation conditions. Each type offers unique advantages and considerations, making it essential to conduct thorough research and consultation before making a decision. By understanding the characteristics and efficiency of different solar panel technologies, homeowners and businesses can make informed choices to maximize the benefits of solar energy for their specific applications.

## 2.3 Installing Solar Panels: Step-by-Step Guide

Installing solar panels involves several key steps to ensure a successful and efficient setup that maximizes the benefits of solar energy. Here's a step-by-step guide to installing solar panels:

**Step 1: Initial Assessment and Planning**

- **Energy Audit**: Conduct an energy audit to understand your current energy consumption and identify areas where energy efficiency improvements can be made. This helps optimize the size and cost-effectiveness of your solar power system.

- **Site Assessment**: Evaluate your property's solar potential by considering factors such as roof orientation, tilt angle, shading from trees or buildings, and available space for solar panel installation. A professional installer can perform a site assessment to determine the optimal placement for solar panels.
- **Permitting and Regulations**: Check local building codes, zoning regulations, and permit requirements for installing solar panels. Obtain necessary permits from local authorities before proceeding with installation.

## Step 2: Designing the Solar Power System

- **System Design**: Work with a qualified solar installer to design a solar power system tailored to your energy needs, site conditions, and budget. The design includes selecting the type and number of solar panels, inverter(s), mounting system, and energy storage (if applicable).
- **Financial Analysis**: Calculate the costs and potential savings of installing solar panels, including upfront expenses, available incentives, financing options, and projected energy savings over time. Determine the return on investment (ROI) and payback period for the system.

## Step 3: Procurement of Equipment

- **Select Solar Panels and Components**: Choose high-quality solar panels, inverters, mounting racks, and other necessary components based on the system design and installer recommendations.

Consider factors such as efficiency, warranty, durability, and compatibility with your system.

- **Order Equipment**: Place orders for solar panels and components from reputable suppliers or through your solar installer. Coordinate delivery and logistics to ensure timely arrival of equipment for installation.

## Step 4: Installation Process

- **Roof Preparation**: Prepare the roof by inspecting its structural integrity and ensuring it can support the weight of solar panels and mounting racks. Repair any roof damage and reinforce if necessary.
- **Mounting Solar Panels**: Install mounting racks or frames on the roof according to the design specifications. Securely attach the racks to the roof rafters or decking to withstand wind and weather conditions.
- **Installing Solar Panels**: Place and secure solar panels onto the mounting racks, ensuring proper alignment and spacing between panels. Connect panels using electrical wiring and conduit to route cables from panels to the inverter(s).
- **Electrical Wiring**: Install inverters, combiner boxes, and other electrical components. Connect DC cables from solar panels to inverters and AC cables from inverters to the electrical panel or grid connection point.
- **Grounding and Safety**: Ensure proper grounding of the solar power system according to electrical codes and safety standards. Install disconnect switches and surge protection devices to safeguard the system and prevent electrical hazards.

## Step 5: Testing and Commissioning

- **System Testing**: Conduct comprehensive testing of the solar power system to verify proper installation, functionality, and performance. Test electrical connections, inverters, and monitoring systems to ensure everything operates correctly.
- **Grid Connection (if applicable)**: If installing a grid-tied system, coordinate with the utility company to connect the solar power system to the grid. Obtain approval and permissions for interconnection and net metering agreements.

## Step 6: Monitoring and Maintenance

- **Monitoring System**: Set up a monitoring system to track the performance and energy production of your solar power system. Monitor solar panel output, inverter efficiency, and overall system performance regularly.
- **Maintenance**: Perform routine maintenance tasks, such as cleaning solar panels, checking electrical connections, and inspecting for shading or debris that could reduce efficiency. Follow manufacturer guidelines for maintenance to ensure optimal system performance and longevity.

## Step 7: Enjoying Solar Energy Benefits

- **Activation**: Activate the solar power system and start generating clean, renewable energy for your home or business. Monitor energy savings, reduce electricity bills, and contribute to environmental sustainability by using solar energy.

- **Educate and Advocate**: Share your experience with solar energy installation and benefits with others. Encourage friends, family, and community members to consider solar power as a viable and sustainable energy solution.

By following these steps and working closely with a qualified solar installer, you can successfully install solar panels and harness the power of the sun to reduce energy costs, increase energy independence, and contribute to a cleaner environment.

## 2.4 Cost and Maintenance Considerations

Installing solar panels involves initial costs and ongoing maintenance considerations that are important for homeowners and businesses to understand when planning to adopt solar energy.

**Initial Costs**

The initial cost of installing solar panels typically includes several components:

1. **Equipment Costs**: The largest portion of the initial cost is the purchase of solar panels, inverters, mounting racks, and other necessary components. The cost varies based on the size and efficiency of the system, quality of equipment, and installation requirements.

2. **Installation Costs**: Professional installation by certified solar installers ensures proper setup and compliance with local building codes and regulations. Installation costs cover labor, equipment setup, electrical wiring, and system testing.
3. **Permitting and Fees**: Obtain necessary permits from local authorities, which may include application fees and inspection costs. These permits ensure that the solar power system meets safety and regulatory standards.
4. **Financing and Incentives**: Various financing options, such as solar loans, leases, and power purchase agreements (PPAs), help offset upfront costs and make solar energy more affordable. Incentives such as tax credits, rebates, and grants from government agencies and utilities further reduce the initial investment.

## Maintenance Considerations

Solar panels are generally low-maintenance but require periodic inspections and upkeep to ensure optimal performance and longevity:

1. **Cleaning**: Regularly clean solar panels to remove dirt, dust, pollen, and debris that can reduce efficiency. Use a soft brush or sponge with mild detergent and water to avoid scratching the panels. Cleaning frequency depends on local climate and environmental factors.
2. **Monitoring**: Install a monitoring system to track energy production and system performance. Monitor solar panel output, inverter efficiency, and overall system operation to identify any issues promptly.
3. **Inspections**: Schedule annual or biannual inspections by a professional solar installer to check electrical connections,

mounting racks, and overall system condition. Inspections help identify potential issues early and ensure compliance with warranty requirements.

4. **Inverter Maintenance**: Inverters are critical components that convert DC electricity from solar panels into usable AC electricity. Follow manufacturer recommendations for inverter maintenance, including cleaning vents, checking connections, and replacing components as needed.

5. **Warranty Coverage**: Solar panels typically come with manufacturer warranties ranging from 10 to 25 years or more, guaranteeing performance and product quality. Inverter warranties often range from 5 to 15 years. Understand warranty terms and coverage for equipment, including repairs and replacements.

6. **Shading and Vegetation**: Monitor for shading from trees, buildings, or other obstructions that can reduce solar panel efficiency. Trim vegetation and remove obstacles that block sunlight from reaching panels.

7. **Snow and Weather Conditions**: In snowy climates, consider snow removal techniques to ensure solar panels continue to generate electricity during winter months. Properly designed mounting racks and panel tilt angles can help minimize snow accumulation.

## Cost Savings and Return on Investment

Despite initial costs and maintenance considerations, solar energy offers significant long-term financial benefits:

1. **Energy Savings**: Generate electricity from sunlight to offset or eliminate electricity bills. Over time, savings on energy costs can

offset initial investment and provide a predictable, stable energy expense.

2. **Return on Investment (ROI)**: Calculate the ROI of installing solar panels by comparing upfront costs with projected energy savings over the system's lifetime. Factors such as financing terms, incentives, and local electricity rates influence ROI calculations.

3. **Property Value**: Increase the resale value of your property with solar panels, which are increasingly attractive to homebuyers seeking energy-efficient homes with reduced operating costs.

4. **Environmental Impact**: Reduce carbon footprint and greenhouse gas emissions by generating clean, renewable energy from solar panels, contributing to environmental sustainability.

In conclusion, understanding the costs and maintenance considerations of installing solar panels is essential for making informed decisions about adopting solar energy. With proper planning, financing, and maintenance, solar panels can provide long-term financial savings, energy independence, and environmental benefits for homeowners and businesses alike.

# Chapter 3: Wind Energy for Residential Use

Wind energy offers a compelling renewable energy solution for residential use, particularly in areas with consistent and strong wind patterns. This chapter explores the fundamentals of wind energy, the types of wind turbines suitable for home use, the installation process, and the benefits and challenges of harnessing wind power at a residential level.

## Fundamentals of Wind Energy

Wind energy is generated by converting the kinetic energy of wind into mechanical power, which is then transformed into electricity through a generator. This process begins when the wind turns the blades of a wind turbine. The rotation of the blades spins a shaft connected to a generator, producing electricity. The amount of electricity generated depends on the wind speed and the size of the turbine.

Wind turbines are most effective in areas with average wind speeds of at least 10 miles per hour. An anemometer is often used to measure wind speed at a prospective site to determine its suitability for wind power generation. Additionally, wind patterns, terrain, and obstructions like buildings or trees must be evaluated to ensure optimal turbine performance.

## Types of Wind Turbines for Residential Use

There are two main types of wind turbines used in residential settings: horizontal-axis wind turbines (HAWTs) and vertical-axis wind turbines (VAWTs).

- **Horizontal-Axis Wind Turbines (HAWTs)**: These are the most common type of wind turbine and have blades similar to airplane propellers. They are typically mounted on a tall tower to capture higher wind speeds and reduce turbulence near the ground. HAWTs are highly efficient and suitable for areas with consistent wind patterns. They require a significant amount of space and can be more visually obtrusive than VAWTs.
- **Vertical-Axis Wind Turbines (VAWTs)**: These turbines have blades that rotate around a vertical axis and come in various designs, such as the Darrieus and Savonius models. VAWTs can capture wind from any direction, making them more versatile in urban or densely built environments where wind direction can be variable. They are generally less efficient than HAWTs but can be installed closer to the ground and require less space.

## Installation Process

Installing a residential wind turbine involves several key steps:

- **Site Assessment**: Conduct a thorough site assessment to evaluate wind speed, direction, and potential obstructions. A professional installer can use tools like anemometers and wind maps to determine the best location for the turbine.
- **Permitting and Regulations**: Obtain necessary permits and approvals from local authorities. Zoning regulations, noise restrictions, and building codes must be considered, as they can affect turbine placement and height.
- **System Design**: Work with a professional installer to design a wind power system tailored to your energy needs and site

conditions. The system design includes selecting the type and size of the turbine, tower height, and electrical components.

- **Foundation and Tower Installation**: Prepare the site by laying a concrete foundation to anchor the turbine tower. Once the foundation is set, the tower is erected, and the turbine is mounted on top.
- **Electrical Connections**: Connect the turbine to an inverter, which converts the generated DC electricity into AC electricity for home use. If the system is grid-tied, the inverter will also manage the flow of electricity between the turbine and the grid.
- **System Testing and Commissioning**: After installation, perform a series of tests to ensure the turbine operates correctly and safely. Monitor the system's performance and make any necessary adjustments.

**Benefits and Challenges**

**Benefits:**

- **Renewable Energy Source**: Wind power is a clean, renewable energy source that reduces reliance on fossil fuels and lowers greenhouse gas emissions.
- **Cost Savings**: Generating your electricity can significantly reduce utility bills and provide long-term financial benefits.
- **Energy Independence**: A residential wind turbine can increase energy independence, especially in areas with frequent power outages.

**Challenges:**

- **Initial Costs**: The upfront costs for purchasing and installing a wind turbine can be high, though various incentives and financing options are available.
- **Maintenance**: Wind turbines require regular maintenance to ensure optimal performance and longevity. This includes inspections, lubrication, and occasional repairs.
- **Aesthetic and Noise Concerns**: Wind turbines can be visually obtrusive and produce noise, which may be a concern for homeowners and neighbors.

In conclusion, wind energy offers a viable renewable energy solution for residential use, particularly in areas with strong and consistent winds. By understanding the types of turbines, the installation process, and the benefits and challenges, homeowners can make informed decisions about adopting wind power to enhance their energy sustainability and independence.

## 3.1 Introduction to Wind Energy

Wind energy, a cornerstone of renewable energy sources, harnesses the natural movement of air to generate electricity. It stands out as a clean, sustainable, and increasingly cost-effective solution to the growing demand for energy, offering a viable alternative to fossil fuels and their associated environmental impacts.

Wind energy is derived from the kinetic energy produced by the movement of air. This kinetic energy is captured by wind turbines, which convert it into mechanical power. The process begins with the wind turning the blades of the turbine. These blades are connected to a rotor, which spins a shaft. This shaft is connected to a generator that converts the mechanical energy into electrical energy. The amount of electricity produced is directly proportional to the wind speed and the size of the turbine, with higher wind speeds and larger turbines generating more power.

The concept of harnessing wind power is not new. Windmills have been used for centuries for tasks such as grinding grain and pumping water. However, the modern wind turbine, designed specifically for electricity generation, has evolved significantly since its inception. Today's turbines are highly efficient and technologically advanced, capable of generating substantial amounts of electricity with minimal environmental impact.

Wind energy's appeal lies in its numerous benefits. Firstly, it is an inexhaustible resource. As long as the wind blows, energy can be harvested. This contrasts starkly with finite fossil fuels, which are depleting rapidly. Wind energy is also environmentally friendly. Unlike coal or natural gas, it produces no greenhouse gas emissions during operation, helping mitigate climate change and reduce air pollution. Additionally, wind farms can be established on agricultural land, allowing for dual land use. Farmers can continue to cultivate crops or graze livestock while simultaneously generating income from wind energy.

The technology behind wind energy has advanced rapidly, making it one of the most cost-effective renewable energy sources. The cost of wind-generated electricity has decreased significantly over the past few decades, making it competitive with traditional fossil fuel-based energy sources. Improvements in turbine design, materials, and manufacturing processes have contributed to this cost reduction, along with economies of scale as wind farms have grown larger.

However, wind energy is not without its challenges. One of the primary issues is variability. The wind is an intermittent resource, meaning that it does not blow consistently. This variability can lead to fluctuations in electricity generation, posing challenges for integrating wind power into the grid. Advances in energy storage technologies, such as batteries, and grid management strategies are helping to address this issue, ensuring a more stable and reliable supply of wind-generated electricity.

Another challenge is the initial cost and infrastructure required for wind turbine installation. Although the long-term savings and environmental benefits are substantial, the upfront investment can be significant. Wind turbines also require maintenance to ensure optimal performance and longevity. This includes regular inspections, lubrication of moving parts, and repairs when necessary.

Public perception and acceptance of wind energy can also pose challenges. While many people support renewable energy in principle, local opposition can arise when new wind farms are proposed, often due to concerns about visual impact, noise, and effects on local wildlife. Effective community engagement and careful site selection are essential to addressing these concerns and gaining public support.

In conclusion, wind energy is a powerful and viable renewable energy source with significant environmental and economic benefits. It harnesses the natural power of the wind to generate clean, sustainable electricity, reducing our reliance on fossil fuels and helping to combat climate change. While there are challenges to overcome, particularly regarding variability and public acceptance, advances in technology and strategic planning are paving the way for a wind-powered future. As part of a diverse and resilient energy mix, wind energy has the potential to play a crucial role in meeting global energy needs sustainably.

## 3.2 Types of Residential Wind Turbines

Wind turbines for residential use come in various designs and sizes, each with specific advantages and applications. The two primary categories are horizontal-axis wind turbines (HAWTs) and vertical-axis wind turbines (VAWTs). Understanding the differences, benefits, and limitations of each type helps homeowners make informed decisions about which turbine best suits their energy needs and site conditions.

**Horizontal-Axis Wind Turbines (HAWTs)**

HAWTs are the most common type of wind turbine. They have blades that rotate around a horizontal axis, similar to a traditional windmill. These turbines are designed to face the wind, with the rotor located at the front of the turbine. Key characteristics include:

- **Design and Functionality**: HAWTs typically have three blades mounted on a tall tower. The height of the tower helps capture stronger and more consistent winds, reducing turbulence caused by ground-level obstructions. The blades are aerodynamically shaped to maximize efficiency, converting more wind energy into electrical power.
- **Efficiency**: HAWTs are known for their high efficiency, often ranging from 35% to 45%. This means they can generate more electricity per unit of wind speed compared to other types of turbines, making them suitable for areas with consistent and strong winds.
- **Installation**: Installing HAWTs usually requires a significant amount of space and a solid foundation to support the tower's

height and weight. The installation process can be complex and may involve significant upfront costs.

- **Maintenance**: HAWTs require regular maintenance, including inspections of the blades, tower, and mechanical components. Maintenance can be challenging due to the height of the tower, often necessitating professional services.

**Vertical-Axis Wind Turbines (VAWTs)**

VAWTs have blades that rotate around a vertical axis, perpendicular to the ground. They are less common than HAWTs but offer unique advantages that make them suitable for specific residential applications. Key characteristics include:

- **Design and Functionality**: VAWTs can capture wind from any direction, making them more versatile in locations with variable wind patterns. They come in various designs, including the Darrieus (eggbeater) and Savonius (s-shaped) models. The Darrieus design is more efficient but requires a higher wind speed to start, while the Savonius design is less efficient but can operate in lower wind speeds.
- **Efficiency**: VAWTs generally have lower efficiency compared to HAWTs, typically ranging from 30% to 40%. However, their ability to capture wind from all directions can offset some of this inefficiency in turbulent environments.
- **Installation**: VAWTs are usually installed closer to the ground and require less space than HAWTs. Their installation is often simpler and less expensive, making them suitable for urban or suburban settings where space is limited.

- **Maintenance**: Maintenance for VAWTs is generally easier due to their lower height. The mechanical components are often located at the base of the turbine, simplifying inspections and repairs.

## Hybrid and Other Types

In addition to HAWTs and VAWTs, hybrid designs combine elements of both types to optimize performance and adaptability. These turbines aim to capture the best features of both horizontal and vertical axis designs, improving efficiency and versatility.

## Factors to Consider

- **Wind Speed and Site Conditions**: Assessing local wind speed and patterns is crucial in selecting the appropriate type of turbine. HAWTs are better suited for areas with strong, consistent winds, while VAWTs can perform well in locations with variable or turbulent wind conditions.
- **Space and Aesthetics**: The available space and aesthetic preferences also play a role in choosing a wind turbine. HAWTs require more space and are often taller, which can be a concern for visual impact. VAWTs are more compact and can blend into the surroundings more easily.
- **Noise and Vibrations**: Consideration of noise and vibrations is important, especially in residential areas. HAWTs can be noisier due to their higher rotational speeds, while VAWTs tend to operate more quietly, making them a better choice for densely populated areas.

- **Budget and Maintenance**: Initial costs, installation complexity, and maintenance requirements vary between HAWTs and VAWTs. It's essential to evaluate the total cost of ownership, including installation, maintenance, and potential energy savings over time.

In conclusion, the choice between horizontal-axis and vertical-axis wind turbines depends on various factors, including wind conditions, space, budget, and aesthetic preferences. Each type offers distinct advantages and challenges, making it crucial for homeowners to thoroughly assess their specific needs and site conditions before deciding on the most suitable wind energy solution for their residence.

## 3.3 Site Assessment and Installation Process

Installing a residential wind turbine involves a thorough site assessment and a detailed installation process to ensure the system's efficiency and longevity. Proper planning and execution are crucial to harnessing wind energy effectively and maximizing the benefits of this renewable resource.

**Site Assessment**

- **Wind Speed and Patterns**: The first step in site assessment is measuring the average wind speed and understanding wind patterns. This can be done using an anemometer to collect wind speed data over time. Ideally, the site should have an average wind speed of at least 10 miles per hour. Wind maps and data from local weather stations can also provide valuable insights into the area's wind resources.

- **Obstructions and Terrain**: Evaluate the site for potential obstructions such as buildings, trees, and hills that can disrupt wind flow and create turbulence. The terrain's elevation and the surrounding landscape significantly influence wind speed and consistency. Open areas, such as fields or hilltops, are typically more suitable for wind turbine installation.

- **Space Requirements**: Determine the available space for the turbine and tower. Horizontal-axis wind turbines (HAWTs) require more space and are usually installed on taller towers, whereas vertical-axis wind turbines (VAWTs) need less space and can be installed closer to the ground. Ensure that there is enough room for the turbine to operate safely without interference from nearby structures.

- **Zoning and Permits**: Check local zoning laws, building codes, and permit requirements. Some areas may have restrictions on the height and placement of wind turbines. Obtaining the necessary permits and complying with local regulations is essential to avoid legal issues and ensure the project's feasibility.

- **Utility Interconnection**: For grid-tied systems, assess the feasibility of connecting the wind turbine to the local power grid. This involves coordinating with the utility company to understand interconnection standards, net metering policies, and any required upgrades to the electrical infrastructure.

## Installation Process

- **System Design**: Collaborate with a professional installer to design the wind energy system. This includes selecting the type and size of the wind turbine, determining the tower height, and planning the electrical connections. The design should be tailored to the site's wind resources, energy needs, and budget.

- **Foundation Preparation**: Prepare the site by laying a concrete foundation to anchor the turbine tower. The foundation must be sturdy enough to support the tower's weight and withstand wind forces. This step involves excavation, pouring concrete, and allowing it to cure properly before proceeding with the tower installation.
- **Tower Installation**: Erect the tower according to the design specifications. Towers can be guyed, monopole, or lattice structures, each with different installation methods and stability features. Secure the tower to the foundation using bolts and ensure it is plumb and level.
- **Turbine Assembly**: Assemble the wind turbine components, including the rotor, blades, and nacelle, which houses the generator and other mechanical parts. The assembly process varies depending on the turbine type and manufacturer. Follow the manufacturer's instructions and guidelines to ensure proper assembly and safety.
- **Electrical Wiring**: Install the electrical wiring to connect the wind turbine to the inverter and the home's electrical system. For grid-tied systems, this includes connecting to the utility grid. Use appropriate wiring and protective devices to ensure safety and compliance with electrical codes. Install disconnect switches, circuit breakers, and surge protectors to safeguard the system from electrical faults.
- **Inverter Installation**: Set up the inverter, which converts the direct current (DC) produced by the wind turbine into alternating current (AC) for home use or grid export. Ensure the inverter is properly sized and compatible with the wind turbine. Mount the

inverter in a location that is easily accessible for maintenance and monitoring.

- **System Testing**: Conduct a series of tests to verify that the wind turbine operates correctly and efficiently. This includes checking the electrical connections, monitoring the turbine's performance, and ensuring the safety systems are functional. Perform initial start-up procedures to confirm that the turbine generates electricity as expected.
- **Commissioning and Inspection**: After successful testing, the system is commissioned and put into regular operation. Schedule a final inspection with local authorities to ensure compliance with all building codes and permit requirements. This step may involve inspections by utility representatives for grid-tied systems.
- **Maintenance Planning**: Develop a maintenance schedule to keep the wind turbine in optimal condition. Routine maintenance tasks include inspecting the blades and tower, lubricating moving parts, checking electrical connections, and monitoring the system's performance. Regular maintenance helps prevent issues and extends the turbine's lifespan.

A thorough site assessment and a well-planned installation process are crucial for the successful deployment of a residential wind turbine. By carefully evaluating wind resources, addressing potential obstructions, complying with regulations, and following a detailed installation plan, homeowners can effectively harness wind energy to reduce their reliance on conventional power sources and contribute to a sustainable future.

# 3.4 Advantages and Limitations of Wind Energy

Wind energy presents numerous benefits and some limitations that must be considered when evaluating its viability as a renewable energy source for residential use. Understanding both sides is crucial for homeowners contemplating the adoption of wind turbines.

**Advantages of Wind Energy**

- **Environmental Benefits**: Wind energy is a clean and renewable energy source that produces no greenhouse gas emissions during operation. By harnessing wind power, homeowners can significantly reduce their carbon footprint and contribute to mitigating climate change. Additionally, wind energy does not produce air pollutants or toxic byproducts, improving local air quality and public health.
- **Sustainability**: Wind is an inexhaustible resource, unlike fossil fuels, which are finite and depleting. As long as the wind blows, energy can be generated, providing a sustainable solution to meet long-term energy needs. This makes wind energy a reliable component of a diversified renewable energy portfolio.
- **Cost Savings**: While the initial investment in wind turbines can be high, the operational costs are relatively low. The wind is free, and once a turbine is installed, it generates electricity at no additional fuel cost. Over time, homeowners can achieve significant savings on their energy bills, particularly in areas with high electricity rates and strong wind resources.
- **Energy Independence**: By generating their electricity, homeowners can reduce their reliance on the grid and conventional power sources. This energy independence provides greater security

against fluctuating energy prices and potential power outages. It also allows for greater control over energy consumption and production.

- **Job Creation and Economic Benefits**: The wind energy industry creates jobs in manufacturing, installation, maintenance, and related sectors. Supporting wind energy can stimulate local economies and provide employment opportunities, contributing to economic development and community resilience.
- **Scalability and Versatility**: Wind turbines can be scaled to fit different energy needs and site conditions. Small wind turbines are suitable for residential use, while larger turbines can power entire communities. This scalability makes wind energy a flexible solution adaptable to various applications and environments.

## Limitations of Wind Energy

- **Intermittency and Variability**: Wind energy is inherently intermittent, as wind does not blow consistently. This variability can lead to fluctuations in electricity generation, posing challenges for integrating wind power into the grid and ensuring a reliable energy supply. Energy storage solutions, such as batteries, and hybrid systems combining wind with other renewable sources can help address this issue.
- **Initial Costs and Financial Barriers**: The upfront costs of purchasing and installing wind turbines can be substantial, including expenses for equipment, site preparation, and permits. While incentives and financing options can offset some of these costs, the financial barrier can be a significant deterrent for many homeowners.

- **Space and Aesthetic Concerns**: Wind turbines require adequate space for installation, which may not be available in densely populated or urban areas. Additionally, some people find wind turbines visually intrusive and may oppose their installation due to concerns about the impact on the landscape and local aesthetics.
- **Noise and Wildlife Impact**: Wind turbines can generate noise during operation, which may be a concern for nearby residents. Advances in turbine design have reduced noise levels, but it remains a consideration. Additionally, wind turbines can pose risks to wildlife, particularly birds and bats. Proper site selection and mitigation measures, such as avoiding migration paths and employing technology to deter wildlife, can help minimize these impacts.
- **Maintenance and Durability**: While wind turbines are generally low maintenance, they require periodic inspections and upkeep to ensure optimal performance and longevity. This includes checking for wear and tear, lubricating moving parts, and repairing any damage. The durability of turbines can also be affected by extreme weather conditions, necessitating robust design and construction.
- **Regulatory and Permitting Challenges**: Navigating the regulatory landscape for wind turbine installation can be complex. Zoning laws, building codes, and permitting requirements vary by location and can impose restrictions on turbine height, placement, and operation. Securing the necessary approvals can be time-consuming and challenging, requiring careful planning and compliance with local regulations.

Wind energy offers significant advantages as a renewable energy source, including environmental benefits, sustainability, cost savings, and energy independence. However, it also presents challenges, such as intermittency, initial costs, space requirements, noise, wildlife impacts, maintenance needs, and regulatory hurdles. By weighing these advantages and limitations, homeowners can make informed decisions about incorporating wind energy into their energy strategy, contributing to a cleaner and more sustainable future.

# Chapter 4: Geothermal Energy Systems

Geothermal energy systems offer a sustainable and efficient solution for residential heating, cooling, and electricity generation by tapping into the Earth's natural heat. This chapter explores the fundamentals of geothermal energy, the types of systems available for residential use, the installation process, and the benefits and challenges of adopting geothermal technology in homes.

## Fundamentals of Geothermal Energy

Geothermal energy harnesses the heat stored beneath the Earth's surface. This heat originates from the Earth's core and the decay of radioactive materials in rocks. The temperature increases with depth, providing a reliable and consistent source of thermal energy. Geothermal systems utilize this heat for various applications, such as heating buildings, providing hot water, and generating electricity.

The key component of a geothermal system is the heat pump, which transfers heat between the ground and the home. In heating mode, the heat pump extracts heat from the ground and distributes it through a ductwork system. In cooling mode, the process is reversed, with heat being extracted from the home and released into the ground.

## Types of Geothermal Systems

There are several types of geothermal systems suitable for residential use, each with specific applications and advantages:

- **Closed-Loop Systems**: These systems circulate a heat-transfer fluid (usually a mixture of water and antifreeze) through a closed loop of pipes buried in the ground. The loops can be installed horizontally, vertically, or in a pond/lake.
- **Horizontal Closed-Loop Systems**: Ideal for homes with ample land, these systems involve burying pipes in trenches about four to six feet deep. They are typically less expensive to install than vertical systems but require more land area.
- **Vertical Closed-Loop Systems**: Suitable for homes with limited land, these systems involve drilling boreholes up to 400 feet deep and inserting pipes vertically. They are more expensive to install but require less land area and are less affected by seasonal temperature variations.
- **Pond/Lake Closed-Loop Systems**: If a body of water is available, pipes can be submerged to exchange heat with the water. This system is cost-effective and efficient but requires proximity to a suitable water source.
- **Open-Loop Systems**: These systems use groundwater from a well or surface water source to transfer heat. Water is pumped through the heat pump and then returned to the ground or discharged to a surface water body. Open-loop systems are highly efficient but depend on an adequate and clean water supply.
- **Hybrid Systems**: These systems combine geothermal with other renewable energy sources, such as solar or wind, to enhance efficiency and reliability. Hybrid systems can optimize energy usage by leveraging multiple sources of renewable energy.

## Installation Process

- **Site Assessment**: Conduct a thorough site assessment to evaluate soil conditions, groundwater availability, and space for loop installation. Soil composition, thermal conductivity, and moisture content are critical factors in determining the system's efficiency.

- **System Design**: Work with a professional installer to design a geothermal system tailored to your energy needs and site conditions. This includes selecting the type of system (closed-loop or open-loop), sizing the heat pump, and planning the layout of the ground loop.
- **Permitting and Regulations**: Obtain necessary permits and approvals from local authorities. Geothermal installations may be subject to regulations regarding drilling, water usage, and environmental impact. Ensuring compliance with local codes is essential.
- **Loop Installation**: Install the ground loop according to the design specifications. This involves trenching or drilling, laying the pipes, and connecting them to the heat pump. For closed-loop systems, ensure the pipes are properly spaced and buried at the correct depth to optimize heat exchange.
- **Heat Pump Installation**: Install the heat pump unit inside the home, typically in the basement or mechanical room. Connect the heat pump to the ground loop and integrate it with the home's existing heating and cooling distribution system.
- **Electrical Connections**: Connect the heat pump to the electrical system and install any necessary controls and monitoring equipment. Ensure all electrical work complies with local codes and standards.
- **System Testing and Commissioning**: After installation, perform a series of tests to verify the system's operation and efficiency. Check for leaks, ensure proper fluid flow, and calibrate the heat pump. Monitor the system's performance to ensure it meets heating and cooling needs.

# Benefits of Geothermal Systems

- **Energy Efficiency**: Geothermal systems are highly efficient, providing three to four units of energy for every unit of electricity consumed. This results in significant energy savings compared to traditional heating and cooling systems.
- **Environmental Impact**: Geothermal energy is a clean and renewable resource, producing no greenhouse gas emissions during operation. It reduces reliance on fossil fuels and helps mitigate climate change.
- **Cost Savings**: Although the initial installation cost can be high, geothermal systems offer long-term savings on energy bills. They also have lower maintenance costs and longer lifespans compared to conventional HVAC systems.
- **Comfort and Reliability**: Geothermal systems provide consistent indoor temperatures and humidity control, enhancing comfort. They operate quietly and have fewer moving parts, resulting in increased reliability and reduced maintenance needs.

# Challenges of Geothermal Systems

- **Initial Costs**: The upfront costs for geothermal system installation can be substantial, including expenses for drilling, equipment, and site preparation. However, incentives and financing options can help offset these costs.
- **Site Suitability**: Not all sites are suitable for geothermal installations. Soil conditions, groundwater availability, and space limitations can affect the feasibility and efficiency of the system.

- **Permitting and Regulatory Hurdles**: Navigating the permitting process and complying with local regulations can be complex and time-consuming. It's essential to work with experienced professionals to ensure all requirements are met.
- **Disruption During Installation**: Installing the ground loop can be disruptive, involving excavation or drilling. Proper planning and coordination can minimize the impact on the property and surrounding area.

In conclusion, geothermal energy systems offer a highly efficient, sustainable, and reliable solution for residential heating and cooling. By understanding the fundamentals, types of systems, installation process, and benefits and challenges, homeowners can make informed decisions about adopting geothermal technology to enhance their energy sustainability and reduce their environmental impact.

## 4.1 Understanding Geothermal Energy

Geothermal energy is a renewable energy source derived from the natural heat of the Earth. This heat originates from the Earth's core, which is composed of molten rock and metal, and from the decay of radioactive materials in the crust. The heat continuously flows outward from the core to the surface, where it can be harnessed for various energy needs. Understanding the principles of geothermal energy and its applications is essential for appreciating its potential as a sustainable energy solution.

## Heat Source and Transfer

The Earth's core generates vast amounts of heat, with temperatures reaching up to 9,000 degrees Fahrenheit. This heat is transferred to the mantle and crust through conduction and convection. The geothermal gradient, which measures the increase in temperature with depth, typically averages about 1.5 degrees Fahrenheit per 100 feet. This consistent heat flow provides a reliable energy source that can be tapped into at various depths.

Geothermal energy can be extracted through several methods, primarily focusing on heat pumps for residential heating and cooling and geothermal power plants for electricity generation. For residential use, geothermal heat pumps (GHPs) are the most common application. GHPs exploit the relatively stable ground temperatures a few feet below the surface to provide efficient heating and cooling.

## Geothermal Heat Pumps

Geothermal heat pumps, also known as ground-source heat pumps, use the Earth's natural heat to regulate indoor temperatures. The system consists of a heat pump, a ground heat exchanger, and a distribution system (ductwork or radiant heating). The ground heat exchanger, a network of pipes buried in the ground, circulates a heat-transfer fluid (water or antifreeze mixture) that absorbs heat from the ground in winter and dissipates heat into the ground in summer.

- **Heating Mode**: In the winter, the heat pump extracts heat from the fluid in the ground heat exchanger and transfers it into the home. The fluid, warmed by the Earth's stable temperatures, passes through the heat pump, where the heat is extracted and distributed through the home's heating system.

- **Cooling Mode**: In the summer, the process reverses. The heat pump extracts heat from the indoor air and transfers it to the fluid in the ground heat exchanger. The heated fluid is then cooled as it circulates through the ground, effectively removing heat from the home and releasing it into the Earth.

## Geothermal Power Plants

For electricity generation, geothermal power plants tap into high-temperature resources found in geothermal reservoirs several miles below the Earth's surface. These reservoirs contain hot water and steam that can be used to drive turbines connected to electricity generators. There are three main types of geothermal power plants:

- **Dry Steam Plants**: These plants use steam directly from the geothermal reservoir to turn turbines. They are the oldest type of geothermal power plant and are primarily used in regions with natural steam resources.
- **Flash Steam Plants**: These plants pull high-pressure hot water from the ground and convert it to steam by reducing the pressure (flashing). The steam then drives the turbines. Flash steam plants are the most common type of geothermal power plant.
- **Binary Cycle Plants**: These plants use a secondary fluid with a lower boiling point than water. The geothermal water heats the secondary fluid, which vaporizes and drives the turbines. Binary cycle plants are used in areas with lower-temperature geothermal resources and have the advantage of being able to operate at lower temperatures.

## Environmental and Economic Benefits

Geothermal energy offers numerous environmental and economic benefits. It is a clean energy source that produces minimal greenhouse gas emissions compared to fossil fuels. By reducing reliance on carbon-intensive energy sources, geothermal energy helps mitigate climate change and improve air quality.

Economically, geothermal energy can provide significant cost savings over time. While the initial installation costs for geothermal systems can be high, the long-term savings on energy bills and maintenance can offset these expenses. Geothermal systems are highly efficient, providing three to four units of energy for every unit of electricity consumed, resulting in lower operational costs.

## Challenges and Considerations

Despite its benefits, geothermal energy also presents some challenges. The initial costs for drilling and installation can be a barrier for many homeowners. Additionally, not all locations are suitable for geothermal installations due to varying geological conditions and the availability of geothermal resources.

Geothermal power plants require specific geological conditions, such as high-temperature reservoirs, which are not available everywhere. The development of these plants can also face environmental concerns, such as the potential for induced seismic activity and land subsidence.

Geothermal energy harnesses the Earth's natural heat to provide a sustainable and efficient energy source for heating, cooling, and electricity generation. By understanding the principles of geothermal heat transfer, the operation of geothermal heat pumps and power plants, and the associated benefits and challenges, homeowners and policymakers can make informed decisions about integrating geothermal energy into their energy strategies. This renewable resource offers a promising path toward reducing greenhouse gas emissions, lowering energy costs, and enhancing energy security.

## 4.2 Types of Geothermal Systems

Geothermal energy systems harness the Earth's natural heat for various applications, from residential heating and cooling to electricity generation. These systems can be broadly categorized into geothermal heat pumps (GHPs) for homes and geothermal power plants for generating electricity. Understanding the different types of geothermal systems is essential for selecting the right solution based on specific needs and site conditions.

### Geothermal Heat Pumps (GHPs)

Geothermal heat pumps, also known as ground-source heat pumps, are the most common type of geothermal system for residential use. They utilize the relatively stable temperatures just below the Earth's surface to provide efficient heating, cooling, and hot water. GHPs consist of a heat pump, a ground heat exchanger, and a distribution system. The primary types of GHPs are:

- **Closed-Loop Systems**: In closed-loop systems, a heat-transfer fluid circulates through a continuous loop of pipes buried in the ground. These systems can be installed in various configurations depending on the available land and specific site conditions.

- **Horizontal Closed-Loop Systems**: These systems are typically used in areas with sufficient land. Pipes are laid out in trenches about four to six feet deep. This configuration is less expensive to install than vertical systems but requires more land area. It is ideal for new construction or large properties.

- **Vertical Closed-Loop Systems**: Suitable for smaller properties or areas with limited land, vertical systems involve drilling boreholes 100 to 400 feet deep. Pipes are inserted vertically into these holes. Although more costly to install due to drilling expenses, vertical systems are more efficient and require less space.

- **Pond/Lake Closed-Loop Systems**: If a property has access to a suitable body of water, pipes can be submerged in the water to exchange heat. This system is cost-effective and highly efficient but requires proximity to a pond or lake with adequate depth and volume.

- **Open-Loop Systems**: Open-loop systems use groundwater or surface water as the heat-transfer medium. Water is drawn from a well or body of water, passed through the heat pump, and then returned to the source or discharged to another location.

- **Groundwater Open-Loop Systems**: These systems use well water for heat exchange. After passing through the heat pump, the water is returned to the ground through a separate discharge well. Open-loop systems are highly efficient but require a reliable and clean water source.

- **Surface Water Open-Loop Systems**: These systems draw water from a lake, river, or pond. The water passes through the heat pump and is then returned to the source. Surface water systems are subject to environmental regulations and may have limited applicability depending on local water conditions and availability.

- **Hybrid Systems**: Hybrid geothermal systems combine geothermal energy with other renewable energy sources, such as solar or wind, to enhance efficiency and reliability. These systems optimize energy use by leveraging multiple energy sources, providing greater flexibility and resilience in various climatic conditions.

## Geothermal Power Plants

Geothermal power plants generate electricity by tapping into high-temperature geothermal resources deep within the Earth's crust. These plants are typically located in regions with significant geothermal activity, such as volcanic areas or tectonic plate boundaries. The main types of geothermal power plants are:

- **Dry Steam Plants**: The oldest type of geothermal power plant, dry steam plants use steam directly from geothermal reservoirs to turn turbines and generate electricity. These plants are limited to areas with natural steam resources, such as geysers and fumaroles. The simplicity of dry steam plants makes them highly efficient, but their application is geographically constrained.
- **Flash Steam Plants**: Flash steam plants are the most common type of geothermal power plant. They utilize high-pressure hot water from geothermal reservoirs. When the water is brought to the surface, the pressure is reduced, causing it to vaporize or "flash" into steam. The steam then drives the turbines to generate electricity. Flash steam plants can operate in a broader range of geothermal environments compared to dry steam plants.
- **Binary Cycle Plants**: Binary cycle plants use a secondary fluid with a lower boiling point than water to generate electricity. Geothermal water heats the secondary fluid, causing it to vaporize

and turn the turbines. The geothermal water and secondary fluid do not mix, and the water is re-injected into the reservoir. Binary cycle plants can operate at lower temperatures, making them suitable for areas with moderate geothermal resources. This flexibility allows binary cycle plants to be deployed in a wider variety of locations.

## Benefits and Challenges

- **Benefits**: Geothermal systems offer several advantages, including high energy efficiency, low operating costs, and minimal environmental impact. They provide consistent heating and cooling, reduce greenhouse gas emissions, and can significantly lower energy bills. Geothermal power plants contribute to a stable and renewable energy supply, reducing reliance on fossil fuels.
- **Challenges**: The initial costs of installing geothermal systems can be high, particularly for drilling and site preparation. Not all locations are suitable for geothermal installations due to varying geological conditions. Additionally, geothermal power plants require specific geothermal resources, which limits their geographical distribution. Environmental concerns, such as potential groundwater contamination and induced seismic activity, must also be addressed.

By understanding the different types of geothermal systems and their specific applications, homeowners and developers can make informed decisions about adopting geothermal technology. This renewable resource holds significant potential for reducing environmental impact, lowering energy costs, and enhancing energy security.

# 4.3 Installation and Maintenance of Geothermal Heat Pumps

Geothermal heat pumps (GHPs) offer a sustainable and efficient way to heat and cool homes by utilizing the Earth's stable underground temperatures. The installation and maintenance of these systems are crucial to ensure optimal performance and longevity. This section provides a detailed overview of the installation process and maintenance requirements for geothermal heat pumps.

## Installation of Geothermal Heat Pumps

- **Site Assessment and Planning**: The first step in installing a GHP system is conducting a thorough site assessment. This involves evaluating the soil composition, moisture content, and thermal conductivity. A professional installer will also assess the available land area to determine the most suitable type of ground loop system—horizontal, vertical, or pond/lake. Based on this assessment, a detailed installation plan is developed, considering factors like local regulations, property layout, and energy needs.
- **Design and Sizing**: Proper sizing of the GHP system is critical for efficiency. The heat pump must be appropriately sized to match the heating and cooling load of the home. Oversized or undersized systems can lead to inefficiencies and increased wear and tear. The design phase also includes planning the layout of the ground loop system, selecting the type of heat pump, and integrating the system with the home's existing ductwork or radiant heating system.
- **Permitting and Approvals**: Installing a GHP system often requires obtaining permits from local authorities. This may involve submitting detailed plans and adhering to specific regulations regarding drilling, excavation, and environmental impact. It's

essential to work with experienced professionals who are familiar with local permitting processes to ensure compliance and avoid delays.

- **Ground Loop Installation**: The ground loop is a crucial component of a GHP system. For horizontal loops, trenches are dug to lay the pipes, which are then covered with soil. Vertical loops require drilling boreholes and inserting pipes vertically. Pond/lake loops involve submerging pipes in a body of water. Proper installation of the ground loop ensures efficient heat transfer and system performance. Care must be taken to avoid damaging the pipes and to ensure they are placed at the correct depth.

- **Heat Pump and Distribution System Installation**: The heat pump unit is installed inside the home, typically in the basement or mechanical room. It is connected to the ground loop and integrated with the home's heating and cooling distribution system. This may involve connecting to existing ductwork or installing a new distribution system. The heat pump must be installed on a stable surface and positioned to allow for proper airflow and maintenance access.

- **Electrical Connections and System Testing**: The heat pump is connected to the electrical system, and all controls and monitoring equipment are installed. The system is then tested to ensure proper operation. This includes checking for leaks in the ground loop, verifying fluid flow, and calibrating the heat pump. The installer will also test the heating and cooling distribution system to ensure it is working correctly and efficiently.

# Maintenance of Geothermal Heat Pumps

- **Regular Inspections**: Regular inspections are essential to maintain the performance and longevity of a GHP system. Homeowners should schedule annual inspections with a qualified technician to check the system's components, including the heat pump, ground loop, and distribution system. Inspections should include checking for leaks, verifying fluid levels, and ensuring proper operation of all controls and sensors.

- **Cleaning and Replacing Filters**: The heat pump's air filters should be cleaned or replaced regularly to maintain efficient airflow and prevent dust and debris from accumulating in the system. Clogged filters can reduce efficiency and increase wear on the heat pump. Depending on the system, filters may need to be cleaned or replaced every few months.

- **Fluid Maintenance**: The heat-transfer fluid in the ground loop should be checked annually for proper levels and concentration. In closed-loop systems, the fluid may need to be replenished or adjusted to ensure optimal heat transfer. Open-loop systems should be checked for water quality and potential scaling or fouling of the heat exchanger.

- **Monitoring System Performance**: Homeowners should monitor their GHP system's performance regularly. This includes keeping an eye on energy bills, indoor temperature consistency, and any unusual noises or system behavior. Sudden changes in performance may indicate a need for maintenance or repair.

- **Addressing Repairs Promptly**: If any issues are detected during inspections or routine monitoring, they should be addressed promptly. Common issues may include leaks in the ground loop, compressor problems, or malfunctioning controls. Timely repairs can prevent minor problems from escalating into more significant issues that could impact the system's efficiency and lifespan.

- **Seasonal Adjustments**: Depending on the climate and seasonal variations, homeowners may need to make adjustments to their GHP system. This can include calibrating thermostats, adjusting fluid levels, and ensuring that the system is optimized for either heating or cooling mode as needed.

The installation and maintenance of geothermal heat pumps are critical to harnessing the full potential of geothermal energy for residential heating and cooling. Proper installation, including thorough site assessment, accurate sizing, and adherence to regulations, ensures optimal performance and efficiency. Regular maintenance, such as inspections, filter cleaning, fluid checks, and timely repairs, helps maintain the system's longevity and reliability. By following these guidelines, homeowners can enjoy the significant benefits of geothermal energy, including reduced energy costs, increased comfort, and a smaller environmental footprint.

## 4.4 Environmental and Economic Benefits

Geothermal energy systems offer numerous environmental and economic benefits that make them an attractive option for residential and commercial use. These benefits stem from the efficient use of the Earth's natural heat to provide sustainable energy solutions for heating, cooling, and electricity generation. Understanding these advantages can help homeowners, businesses, and policymakers make informed decisions about investing in geothermal technologies.

# Environmental Benefits

- **Reduced Greenhouse Gas Emissions**: One of the most significant environmental benefits of geothermal energy is its potential to reduce greenhouse gas emissions. Unlike fossil fuel-based energy sources, geothermal systems produce minimal carbon dioxide and other greenhouse gases. This reduction in emissions helps mitigate climate change and improves air quality, contributing to a healthier environment.

- **Sustainable and Renewable Energy Source**: Geothermal energy is a renewable resource, meaning it is sustainable and can be replenished naturally. The Earth continuously produces heat from its core, providing a virtually inexhaustible energy supply. This sustainability ensures that geothermal energy can be a long-term solution for reducing dependence on finite fossil fuels.

- **Minimal Land and Water Usage**: Geothermal systems have a relatively small land footprint compared to other renewable energy sources like solar and wind farms. Once installed, the ground loops for geothermal heat pumps or the infrastructure for geothermal power plants occupy minimal surface area, preserving land for other uses. Additionally, closed-loop geothermal systems use a small amount of water, reducing the strain on water resources. Open-loop systems may use more water, but they often return it to the source with minimal impact.

- **Low Environmental Impact**: Geothermal systems have a low environmental impact during operation. They operate quietly and produce no air pollution, making them suitable for residential areas and urban environments. Additionally, geothermal power plants typically have lower environmental impacts compared to traditional power plants, as they do not require fuel combustion or significant resource extraction.

# Economic Benefits

- **Energy Cost Savings**: One of the most compelling economic benefits of geothermal energy systems is the potential for significant energy cost savings. Geothermal heat pumps are highly efficient, providing three to four units of energy for every unit of electricity consumed. This efficiency translates into lower heating and cooling costs for homeowners and businesses. Over time, these savings can offset the initial installation costs of geothermal systems.

- **Stable Energy Prices**: Geothermal energy provides a stable and predictable source of energy, helping to shield consumers from the volatility of fossil fuel prices. Because geothermal systems rely on the Earth's constant temperatures, they are less affected by fluctuations in fuel supply and demand. This stability can lead to more predictable energy costs and better financial planning for households and businesses.

- **Long Lifespan and Low Maintenance Costs**: Geothermal systems are known for their durability and longevity. The underground components, such as the ground loops, can last for 50 years or more, while the heat pump units typically have a lifespan of 20 to 25 years. This longevity reduces the need for frequent replacements and associated costs. Moreover, geothermal systems have fewer moving parts compared to traditional heating and cooling systems, resulting in lower maintenance requirements and costs.

- **Increased Property Value**: Homes and buildings equipped with geothermal systems often see an increase in property value. Prospective buyers are increasingly looking for energy-efficient and environmentally friendly features, and geothermal systems can be a strong selling point. The promise of lower energy bills and reduced environmental impact can make properties with geothermal installations more attractive in the real estate market.

- **Government Incentives and Rebates**: Many governments offer incentives, rebates, and tax credits to encourage the adoption of geothermal energy systems. These financial incentives can significantly reduce the upfront costs of installation, making geothermal systems more affordable for homeowners and businesses. Additionally, some regions offer grants and low-interest loans to support geothermal projects, further enhancing their economic viability.

Geothermal energy systems present a compelling case for both environmental and economic benefits. Environmentally, they contribute to reduced greenhouse gas emissions, sustainable energy production, minimal land and water usage, and low operational impact. Economically, they offer significant energy cost savings, stable energy prices, long system lifespan, low maintenance costs, increased property value, and access to government incentives. By understanding and leveraging these benefits, individuals and organizations can make informed decisions about investing in geothermal technology, contributing to a more sustainable and economically resilient future.

# Chapter 5: Biomass and Bioenergy

Biomass and bioenergy represent renewable energy sources derived from organic materials, offering versatile solutions for heat, electricity generation, and transportation fuels. This chapter explores the fundamentals of biomass and bioenergy, various types of biomass feedstocks, conversion technologies, environmental considerations, and their role in sustainable energy production.

## Fundamentals of Biomass and Bioenergy

Biomass refers to organic materials derived from plants and animals that can be used as a renewable energy source. These materials include wood, agricultural residues, energy crops (such as switchgrass and miscanthus), algae, and organic waste. Biomass can be converted into bioenergy through various processes, including combustion, gasification, fermentation, and biochemical conversion.

Bioenergy encompasses different forms of energy derived from biomass, including heat, electricity, and biofuels. It provides a sustainable alternative to fossil fuels by utilizing organic materials that can be replenished through natural processes. Bioenergy systems contribute to reducing greenhouse gas emissions, enhancing energy security, and promoting rural development through biomass production and utilization.

## Types of Biomass Feedstocks

- **Woody Biomass**: Wood and wood residues from forestry and logging operations are common sources of biomass feedstock. Woody biomass can be used for heat and electricity generation through direct combustion in boilers and biomass power plants. It can also be converted into biofuels such as wood pellets and wood chips for residential heating and industrial applications.
- **Agricultural Residues**: Agricultural residues, including crop residues (such as corn stover and wheat straw) and animal manure, are abundant sources of biomass. These residues can be used for bioenergy production through combustion, gasification, and biochemical processes. They provide an opportunity to utilize agricultural waste products for energy generation while reducing greenhouse gas emissions from decomposition.
- **Energy Crops**: Dedicated energy crops, such as switchgrass, miscanthus, and willow, are grown specifically for bioenergy production. These crops offer high biomass yields and can be cultivated on marginal lands unsuitable for food crops. Energy crops are used for biofuel production, biogas generation, and as feedstocks for biochemical conversion processes.
- **Organic Waste**: Organic waste materials, including municipal solid waste, food waste, and sewage sludge, can be processed into biogas through anaerobic digestion. Biogas, primarily composed of methane and carbon dioxide, can be used for electricity generation, heat production, and as a renewable transportation fuel. Anaerobic digestion not only produces energy but also reduces the volume of organic waste sent to landfills, mitigating methane emissions.


## Biomass Conversion Technologies

- **Combustion**: Biomass combustion involves burning organic materials to produce heat. Direct combustion is used in residential heating systems, biomass boilers, and biomass power plants to generate steam for electricity generation. Advanced combustion technologies, such as fluidized bed combustion and biomass gasification, improve efficiency and reduce emissions.

- **Gasification**: Biomass gasification converts solid biomass into a synthetic gas (syngas) containing carbon monoxide, hydrogen, and methane. Syngas can be combusted directly for heat and power generation or further processed into liquid biofuels through Fischer-Tropsch synthesis. Gasification technologies enable efficient utilization of biomass and can be integrated with combined heat and power (CHP) systems.

- **Biochemical Conversion**: Biochemical conversion processes, such as fermentation and anaerobic digestion, use enzymes and microorganisms to break down biomass into sugars, alcohol, and organic acids. These bio-based products can be refined into biofuels, including ethanol, biodiesel, and biogas. Biochemical conversion technologies offer versatility in producing liquid fuels suitable for transportation and stationary applications.

## Environmental Considerations

Biomass and bioenergy systems offer environmental benefits, such as reducing greenhouse gas emissions compared to fossil fuels. Biomass combustion and biofuels are considered carbon-neutral because the carbon dioxide released during combustion is offset by the carbon dioxide absorbed during biomass growth. However, biomass production and harvesting practices must be sustainable to maintain soil fertility, biodiversity, and ecosystem services.

**Role in Sustainable Energy Production**

Biomass and bioenergy play a vital role in sustainable energy production by diversifying the energy mix, reducing dependence on fossil fuels, and mitigating climate change. They contribute to rural economic development by creating jobs in biomass production, harvesting, and processing industries. Bioenergy systems enhance energy security by utilizing locally available biomass resources, reducing reliance on imported fuels, and promoting decentralized energy generation.

Biomass and bioenergy represent versatile and sustainable solutions for meeting energy demands while mitigating environmental impacts. By utilizing organic materials such as wood residues, agricultural waste, energy crops, and organic waste, biomass can be converted into heat, electricity, and biofuels through various conversion technologies. Biomass and bioenergy systems contribute to reducing greenhouse gas emissions, enhancing energy security, and promoting rural development. Understanding the fundamentals, types of feedstocks, conversion technologies, environmental considerations, and their role in sustainable energy production is essential for advancing bioenergy as a renewable energy source in the global energy transition.

## 5.1 Basics of Biomass Energy

Biomass energy refers to the use of organic materials derived from plants and animals to produce heat, electricity, and biofuels. It is a renewable energy source because organic materials can be replenished through natural processes, such as photosynthesis and organic waste decomposition. Biomass energy has been utilized for centuries, primarily through traditional biomass fuels like wood, dung, and crop residues for heating and cooking purposes. In modern times, biomass energy encompasses a broader range of technologies and applications, contributing to sustainable energy production and reducing dependence on fossil fuels.

**Types of Biomass Feedstocks**

Biomass feedstocks include a variety of organic materials that can be converted into energy through different processes:

- **Woody Biomass**: Wood and wood residues from forestry and logging operations are significant sources of biomass feedstock. They can be used for direct combustion in stoves, boilers, and biomass power plants to generate heat and electricity. Wood pellets and wood chips are also used as biofuels for residential heating and industrial applications.
- **Agricultural Residues**: Crop residues such as corn stover, wheat straw, and rice husks are abundant sources of biomass. These residues are leftover from agricultural activities and can be used for bioenergy production through combustion, gasification, and biochemical processes. Agricultural residues offer a sustainable way to utilize waste materials while reducing greenhouse gas emissions from open burning or decomposition.
- **Energy Crops**: Dedicated energy crops, such as switchgrass, miscanthus, and willow, are grown specifically for bioenergy production. These crops have high biomass yields and can be cultivated on marginal lands unsuitable for food crops. Energy crops are used for biofuel production, biogas generation, and as feedstocks for biochemical conversion processes like ethanol production.

- **Organic Waste**: Organic waste materials, including municipal solid waste, food waste, and sewage sludge, can be processed into biogas through anaerobic digestion. Anaerobic digestion breaks down organic matter in the absence of oxygen, producing biogas primarily composed of methane and carbon dioxide. Biogas can be used for electricity generation, heat production, and as a renewable transportation fuel, providing an environmentally friendly alternative to fossil fuels.

**Conversion Technologies**

Biomass can be converted into energy through several technological processes:

- **Combustion**: Biomass combustion involves burning organic materials to produce heat. Direct combustion is used in residential heating systems, biomass boilers, and biomass power plants to generate steam for electricity generation. Advanced combustion technologies, such as fluidized bed combustion, improve efficiency and reduce emissions by optimizing the combustion process.
- **Gasification**: Biomass gasification converts solid biomass into a synthetic gas (syngas) containing carbon monoxide, hydrogen, and methane. Syngas can be combusted directly for heat and power generation or further processed into liquid biofuels through Fischer-Tropsch synthesis. Gasification technologies enable efficient utilization of biomass and can be integrated with combined heat and power (CHP) systems to maximize energy efficiency.
- **Biochemical Conversion**: Biochemical conversion processes use enzymes and microorganisms to break down biomass into sugars,

alcohols, and organic acids. These bio-based products can be refined into biofuels, including ethanol, biodiesel, and biogas. Biochemical conversion technologies offer flexibility in producing liquid fuels suitable for transportation and stationary applications, contributing to reducing greenhouse gas emissions and enhancing energy security.

## Environmental Considerations

Biomass energy systems offer environmental benefits, such as reducing greenhouse gas emissions compared to fossil fuels. Biomass combustion and biofuels are considered carbon-neutral because the carbon dioxide released during combustion is offset by the carbon dioxide absorbed during biomass growth. However, sustainable biomass production and harvesting practices are crucial to maintaining soil fertility, biodiversity, and ecosystem services. Proper management of biomass resources ensures that bioenergy remains a sustainable and environmentally friendly energy option.

## Role in Sustainable Energy Production

Biomass energy plays a critical role in sustainable energy production by diversifying the energy mix, reducing dependence on fossil fuels, and mitigating climate change. It contributes to rural economic development by creating jobs in biomass production, harvesting, and processing industries. Biomass energy systems enhance energy security by utilizing locally available biomass resources, reducing reliance on imported fuels, and promoting decentralized energy generation. By harnessing organic materials for heat, electricity, and biofuels, biomass energy supports a cleaner and more sustainable energy future.

Biomass energy is a versatile and renewable energy source derived from organic materials like wood, agricultural residues, energy crops, and organic waste. Through combustion, gasification, and biochemical conversion technologies, biomass can be converted into heat, electricity, and biofuels, offering sustainable alternatives to fossil fuels. Biomass energy systems provide environmental benefits by reducing greenhouse gas emissions and promoting carbon neutrality. They play a crucial role in sustainable energy production by diversifying the energy mix, enhancing energy security, and supporting rural development. Understanding the basics of biomass energy and its applications is essential for advancing bioenergy as a viable solution in the global transition towards a cleaner and more sustainable energy future.

## 5.2 Types of Biomass and Their Applications

Biomass encompasses a diverse range of organic materials derived from plants and animals that can be converted into energy through various technological processes. Each type of biomass offers unique advantages and applications, contributing to sustainable energy production and reducing dependence on fossil fuels.

### 1. Woody Biomass

Description: Woody biomass consists of wood and wood residues obtained from forestry and logging operations, as well as urban wood waste and construction debris.

**Applications:**

- **Heat and Electricity Generation**: Woody biomass is primarily used for direct combustion in stoves, residential boilers, and biomass power plants. Combustion processes generate heat by burning wood chips, pellets, or logs, which can be used for space heating and water heating in residential, commercial, and industrial settings.
- **Biofuels**: Wood residues can be processed into wood pellets and briquettes, which are used as biofuels in residential and industrial boilers. These biomass fuels offer a renewable alternative to fossil fuels for heating applications, reducing greenhouse gas emissions and promoting sustainable energy practices.
- **Combined Heat and Power (CHP)**: Advanced combustion technologies, such as fluidized bed combustion and biomass gasification, enable the efficient conversion of woody biomass into electricity and heat. Biomass CHP systems maximize energy efficiency by generating both electricity and usable heat from the combustion process, providing a decentralized energy solution for communities and industries.

## 2. Agricultural Residues

Description: Agricultural residues include crop residues, such as corn stover, wheat straw, rice husks, and sugarcane bagasse, left over from agricultural production processes.

**Applications:**

- **Bioenergy Production**: Agricultural residues are valuable feedstocks for bioenergy production through combustion, gasification, and biochemical processes. Combustion of crop residues in biomass boilers and power plants generates heat and electricity, contributing to renewable energy generation and reducing dependence on fossil fuels.
- **Biofuels**: Crop residues can be converted into biofuels, such as ethanol and biodiesel, through biochemical conversion processes. Ethanol production from sugarcane bagasse and corn stover offers a sustainable alternative to gasoline, supporting the transportation sector's transition to renewable energy sources and reducing carbon emissions.
- **Soil Amendment**: Agricultural residues, when returned to the soil through mulching or incorporation, improve soil fertility and structure. Crop residues contribute to soil organic matter content, enhancing nutrient cycling and water retention, thereby promoting sustainable agricultural practices and ecosystem health.

### 3. Energy Crops

Description: Energy crops are dedicated crops grown specifically for bioenergy production, such as switchgrass, miscanthus, willow, and hybrid poplar.

**Applications:**

- **Biofuel Production**: Energy crops offer high biomass yields and can be cultivated on marginal lands unsuitable for food crops. Switchgrass and miscanthus are perennial grasses that can be harvested and processed into biofuels, such as cellulosic ethanol and biogas, supporting the development of renewable transportation fuels and reducing greenhouse gas emissions from conventional gasoline and diesel.
- **Biogas Generation**: Energy crops, including maize and sorghum, are used as feedstocks for biogas production through anaerobic digestion. Biogas is composed of methane and carbon dioxide and can be combusted to generate heat and electricity or refined into renewable natural gas for use in vehicles and industrial applications, promoting sustainable energy practices and reducing methane emissions from organic waste decomposition.

## 4. Organic Waste

Description: Organic waste materials, including municipal solid waste (MSW), food waste, sewage sludge, and livestock manure, are valuable biomass feedstocks for bioenergy production.

**Applications:**

- **Biogas Production**: Anaerobic digestion of organic waste materials converts biodegradable components into biogas, primarily composed of methane and carbon dioxide. Biogas can be utilized for heat and power generation, replacing fossil fuels and reducing greenhouse gas emissions from organic waste disposal.

- **Bioenergy Recovery**: Municipal solid waste (MSW) and food waste can be processed into refuse-derived fuel (RDF) pellets or used as feedstocks for waste-to-energy (WTE) facilities. WTE technologies combust organic waste materials to generate heat and electricity, providing sustainable energy solutions and diverting waste from landfills.

- **Nutrient Recycling**: Organic waste materials, such as livestock manure and sewage sludge, contain valuable nutrients like nitrogen and phosphorus. Anaerobic digestion and composting processes convert organic waste into biofertilizers, promoting soil health and reducing reliance on synthetic fertilizers in agriculture.

Biomass encompasses a diverse range of organic materials that can be converted into renewable energy through combustion, gasification, biochemical conversion, and anaerobic digestion technologies. Each type of biomass offers unique advantages and applications, including heat and electricity generation, biofuel production, biogas generation, and soil amendment. By utilizing biomass feedstocks such as woody biomass, agricultural residues, energy crops, and organic waste, communities and industries can promote sustainable energy practices, reduce greenhouse gas emissions, and enhance energy security. Understanding the types of biomass and their applications is essential for advancing bioenergy as a viable solution in the global transition towards a cleaner and more sustainable energy future.

# 5.3 Home Biomass Systems: Setup and Management

Setting up and managing home biomass systems involves careful planning, installation, and ongoing maintenance to ensure efficient and effective operation. These systems utilize biomass fuels such as wood pellets, logs, or agricultural residues to provide heat and, in some cases, electricity. Here's a detailed overview of how to set up and manage home biomass systems:

## Planning and Installation

### 1. System Selection:

- **Fuel Type**: Choose the fittingbiomass fuel based on availability, cost, and system compatibility. Common fuels include wood pellets, wood chips, and logs.
- **System Type**: Select the type of biomass system based on heating needs and available space. Options include biomass boilers, stoves, and combined heat and power (CHP) systems.

### 2. Sizing and Design:

- **Heat Demand**: Calculate the heating demand of your home to determine the appropriate size of the biomass system. Consider factors such as insulation levels, building size, and climate conditions.

- **Integration**: Plan how the biomass system will integrate with existing heating systems, such as radiators or underfloor heating. Ensure compatibility and efficiency in heat distribution.

## 3. Installation Process:

- **Professional Installation**: Hire qualified installers experienced in biomass systems to ensure proper installation and compliance with local building codes and safety regulations.
- **Fuel Storage**: Install a suitable storage system for biomass fuel, such as a hopper or silo, located near the biomass system for easy refueling.

## Operating and Managing Biomass Systems

## 1. Fuel Management:

- **Fuel Quality**: Use high-quality biomass fuel to ensure efficient combustion and minimize emissions. Follow manufacturer recommendations for fuel type and moisture content.
- **Fuel Storage**: Keep biomass fuel dry and protected from moisture to maintain its quality and combustion efficiency.
- **Regular Refueling**: Monitor fuel levels and refill the storage system as needed to maintain continuous operation during colder months.

## 2. System Maintenance:

- **Cleaning and Inspections**: Regularly clean combustion chambers, flues, and heat exchangers to remove ash and soot buildup. Schedule annual inspections by a qualified technician to check for leaks, assess system efficiency, and perform necessary maintenance.
- **Component Check**: Inspect and replace worn-out components, such as gaskets and seals, to ensure optimal performance and safety of the biomass system.

## 3. Operational Considerations:

- **Startup and Shutdown**: Follow manufacturer instructions for proper startup and shutdown procedures to avoid operational issues and ensure safe operation.
- **Temperature Control**: Use programmable thermostats or control systems to regulate heating temperatures and optimize energy efficiency based on heating demand and outdoor temperatures.

## 4. Emissions and Environmental Impact:

- **Emissions Monitoring**: Monitor emissions from the biomass system to ensure compliance with environmental regulations and minimize air pollutants.
- **Environmental Practices**: Practice responsible fuel sourcing and management to minimize environmental impact, such as sourcing

sustainable biomass fuels and properly disposing of ash and residues.

**Benefits of Home Biomass Systems**

- **Cost Savings**: Biomass systems can offer cost savings compared to conventional heating fuels over the long term, especially in regions with abundant biomass resources.
- **Energy Independence**: Using locally sourced biomass fuels reduces dependence on imported fossil fuels, enhancing energy security and resilience.
- **Environmental Benefits**: Biomass systems can reduce greenhouse gas emissions compared to fossil fuels, contributing to climate change mitigation and environmental sustainability.
- **Renewable Energy**: Biomass is a renewable energy source, utilizing organic materials that can be replenished through natural processes.

Home biomass systems provide a sustainable and efficient heating solution for residential properties, utilizing biomass fuels such as wood pellets or logs to generate heat and, in some cases, electricity. Proper planning, installation, and management are essential to maximize the benefits of biomass systems, including cost savings, energy independence, and environmental sustainability. By following guidelines for fuel management, system maintenance, and operational best practices, homeowners can ensure reliable and efficient operation of their biomass heating systems, contributing to a cleaner and more sustainable energy future.

## 5.4 Sustainability and Environmental Impact

Sustainability and environmental impact are crucial considerations in the context of biomass energy systems, which utilize organic materials to produce heat, electricity, and biofuels. Understanding these factors is essential for evaluating the overall benefits and challenges associated with biomass as a renewable energy source.

### Sustainability of Biomass Energy

Biomass energy is considered sustainable when organic materials used for energy production are sourced responsibly and managed in a way that maintains their availability for future generations. Key aspects of biomass sustainability include:

1. **Renewable Resource**: Biomass is derived from organic materials such as wood, agricultural residues, energy crops, and organic waste, which can be replenished through natural processes over time. Unlike fossil fuels, which are finite and non-renewable, biomass offers a continuous supply of energy as long as sustainable harvesting practices are followed.
2. **Carbon Neutrality**: Biomass combustion and biofuels are often considered carbon-neutral because the carbon dioxide ($CO_2$) released during combustion is part of the carbon cycle. Plants absorb $CO_2$ from the atmosphere through photosynthesis, and when biomass is burned or converted into biofuels, this $CO_2$ is released back into the atmosphere. As long as new biomass growth absorbs an equivalent amount of $CO_2$, biomass energy does not contribute to a net increase in atmospheric $CO_2$ levels.

3. **Land Use and Biodiversity**: Sustainable biomass production considers land use practices that do not compete with food production or lead to deforestation or habitat destruction. Energy crops like switchgrass and miscanthus can be grown on marginal lands, reducing pressure on prime agricultural land and preserving biodiversity. Managed forestry practices ensure that wood biomass harvesting maintains forest health and biodiversity.
4. **Lifecycle Assessment**: Lifecycle assessments (LCAs) evaluate the environmental impacts of biomass energy systems throughout their entire lifecycle, from feedstock production and harvesting to processing, transportation, combustion, and disposal of residues. LCAs consider factors such as greenhouse gas emissions, air and water pollution, energy efficiency, and resource depletion to assess the overall environmental footprint of biomass energy compared to fossil fuels.

## Environmental Impact of Biomass Energy

While biomass energy offers potential environmental benefits, it also poses certain environmental challenges that need careful consideration and management:

1. **Air Quality**: Biomass combustion releases pollutants such as particulate matter (PM), nitrogen oxides (NOx), sulfur dioxide (SO2), and volatile organic compounds (VOCs). Modern biomass combustion technologies, such as advanced boilers and emissions control systems, can reduce emissions and improve air quality compared to traditional wood stoves and open-burning practices.
2. **Water Use and Quality**: Biomass cultivation and processing may require water for irrigation, biomass washing, or cooling systems.

Efficient water management practices are essential to minimize water use and prevent contamination of water resources from agricultural runoff or biomass processing effluents.

3. **Land and Soil Impacts**: Intensive biomass cultivation can lead to soil erosion, nutrient depletion, and changes in soil structure and composition. Sustainable land management practices, such as crop rotation, cover cropping, and conservation tillage, help mitigate these impacts and maintain soil fertility and productivity.

4. **Waste Management**: Biomass energy systems generate ash and residues from combustion or biochemical conversion processes. Proper disposal or utilization of biomass residues, such as ash recycling in agriculture or forestry, minimizes waste and enhances resource efficiency.

**Promoting Sustainable Biomass Energy**

To promote sustainable biomass energy, stakeholders can adopt the following strategies:

- **Policy Support**: Implement policies and regulations that promote sustainable biomass sourcing, land use practices, emissions standards, and waste management strategies.
- **Technology Innovation**: Invest in research and development of advanced biomass conversion technologies, energy efficiency improvements, and emissions control technologies to enhance environmental performance.
- **Certification and Standards**: Establish certification schemes and sustainability standards for biomass feedstocks and energy systems to ensure compliance with environmental criteria and promote transparency in biomass supply chains.

- **Public Awareness and Education**: Raise awareness among consumers, policymakers, and industries about the benefits and challenges of biomass energy, emphasizing the importance of sustainable practices and responsible energy use.

Sustainability and environmental impact are central considerations in the development and deployment of biomass energy systems. By promoting sustainable biomass sourcing, adopting efficient technologies, and mitigating environmental impacts, biomass energy can play a significant role in reducing greenhouse gas emissions, enhancing energy security, and promoting rural development. Continued efforts to improve biomass sustainability through policy support, technological innovation, and public engagement are essential for maximizing the environmental benefits of biomass energy and advancing towards a more sustainable energy future.

# Chapter 6: Energy Storage Solutions

Energy storage solutions play a pivotal role in modern energy systems by addressing the intermittent nature of renewable energy sources like solar and wind. These technologies store excess energy when supply exceeds demand and release it when demand is higher than supply, ensuring grid stability, enhancing energy reliability, and supporting the integration of renewable energy into the electricity grid.

## Importance of Energy Storage

Energy storage systems are critical for balancing supply and demand dynamics in the electricity grid, particularly as the share of renewable energy sources increases. Solar and wind energy generation fluctuates based on weather conditions, making it challenging to match supply with variable demand patterns throughout the day. Energy storage solutions enable the capture and storage of surplus renewable energy during periods of high generation for later use during peak demand periods or when renewable generation is low.

## Types of Energy Storage Technologies

### 1. Battery Storage Systems:

- **Lithium-ion Batteries**: Widely used for their high energy density, efficiency, and scalability. Lithium-ion batteries are suitable for both grid-scale and residential applications, providing short-duration energy storage solutions.

- **Flow Batteries**: Utilize chemical components stored outside the battery cell, offering flexibility in energy capacity and longer-duration storage capabilities. Flow batteries are advantageous for grid-scale applications requiring extended discharge periods.

## 2. Pumped Hydro Storage:

Mechanism: Stores energy by pumping water to an elevated reservoir during periods of low demand and releasing it through turbines to generate electricity during peak demand. Pumped hydro storage facilities are known for their large-scale energy storage capacity and long operational life spans.

## 3. Thermal Energy Storage:

Concept: Stores energy in the form of heat using materials such as molten salts or phase-change materials. Thermal energy storage systems can be integrated with concentrating solar power plants to provide dispatchable electricity generation even after sunset or during cloudy weather.

## 4. Compressed Air Energy Storage (CAES):

Process: Stores energy by compressing air into underground caverns or storage tanks during low-demand periods and releasing it through turbines to generate electricity during high-demand periods. CAES systems offer large-scale energy storage capabilities and can be coupled with renewable energy sources like wind farms.

**Applications of Energy Storage Solutions**

**1. Grid Stability and Reliability:**

Energy storage systems enhance grid stability by providing frequency regulation, voltage support, and grid balancing services. They help mitigate grid imbalances caused by fluctuating renewable energy generation and sudden changes in electricity demand.

**2. Renewable Energy Integration:**

Energy storage facilitates the integration of intermittent renewable energy sources, such as solar and wind, into the electricity grid. By storing excess renewable energy during periods of high generation and releasing it during periods of high demand or low generation, energy storage supports a more reliable and sustainable energy supply.

**3. Peak Shaving and Demand Response:**

Energy storage systems reduce electricity costs by storing energy during off-peak hours when electricity prices are low and supplying it during peak demand periods when prices are high. This practice, known as peak shaving, helps consumers and utilities manage electricity costs more effectively.

## 4. Emergency Backup Power:

Residential and commercial energy storage systems provide backup power during grid outages or emergencies, ensuring the continuity of critical services and reducing reliance on diesel generators or grid-supplied electricity.

## Future Trends and Challenges

1. **Cost Reduction**: Continued advancements in energy storage technologies, manufacturing processes, and economies of scale are expected to drive down costs, making energy storage solutions more economically viable for widespread deployment.
2. **Regulatory Support**: Policies and regulations supporting energy storage deployment, grid integration, and market participation are crucial for accelerating the adoption of energy storage solutions and maximizing their benefits for grid reliability and renewable energy integration.
3. **Technological Innovation**: Research and development in next-generation energy storage technologies, such as solid-state batteries, advanced flow batteries, and hydrogen-based storage systems, hold promise for further enhancing energy storage efficiency, capacity, and sustainability.

Energy storage solutions are essential for addressing the variability and intermittency of renewable energy sources and enhancing the reliability, flexibility, and sustainability of modern energy systems. By enabling the efficient storage and utilization of surplus renewable energy, energy storage technologies play a pivotal role in achieving a cleaner and more resilient energy future. Continued advancements in energy storage technologies, coupled with supportive policies and regulatory frameworks, will be key to unlocking the full potential of energy storage solutions in advancing towards a more sustainable and low-carbon energy system.

## 6.1 Importance of Energy Storage

Energy storage is increasingly recognized as a crucial component of modern energy systems, essential for ensuring grid stability, enhancing energy reliability, and facilitating the integration of renewable energy sources. This chapter explores the significance of energy storage technologies in today's energy landscape.

### Grid Stability and Reliability

One of the primary roles of energy storage is to enhance grid stability and reliability. Electricity demand fluctuates throughout the day, and traditional power plants adjust their output to match these fluctuations. However, renewable energy sources such as solar and wind are intermittent, depending on weather conditions. Energy storage systems address this variability by storing excess energy when renewable generation exceeds demand and releasing stored energy during periods of high demand or low generation. This capability helps balance supply and demand in real time, stabilizing the grid and minimizing disruptions.

## Integration of Renewable Energy

Energy storage solutions play a critical role in the integration of renewable energy sources into the electricity grid. Solar and wind power generation can be unpredictable and often do not coincide with peak electricity demand periods. Energy storage systems enable the capture and storage of surplus renewable energy during times of high generation for later use when renewable generation is low or when electricity demand peaks. By smoothing out fluctuations in renewable energy output, energy storage facilitates a more reliable and sustainable energy supply, reducing the need for backup fossil fuel power plants and supporting the transition to a low-carbon energy system.

## Peak Shaving and Demand Management

Energy storage systems contribute to peak shaving and demand management strategies, which optimize electricity consumption and reduce costs for utilities and consumers. During off-peak hours when electricity demand is low, energy storage systems can store excess energy. This stored energy is then discharged during peak demand periods when electricity prices are higher, reducing overall electricity costs and alleviating strain on the grid. By shifting electricity consumption patterns, energy storage supports more efficient use of electricity resources and enhances grid efficiency.

## Grid Resilience and Emergency Backup

Energy storage systems provide grid resilience by offering backup power during grid outages, natural disasters, or emergencies. Distributed energy storage solutions installed at residential, commercial, and industrial sites can operate independently of the main grid, ensuring continuity of electricity supply for critical services and reducing reliance on diesel generators or other backup sources. Grid-connected energy storage systems can also support grid stability and recovery efforts following disruptions, contributing to overall grid resilience and reliability.

## Economic Benefits and Energy Security

Energy storage technologies offer significant economic benefits by reducing overall electricity costs, enhancing energy market competitiveness, and supporting local energy independence. By optimizing energy use, reducing peak demand charges, and integrating renewable energy sources, energy storage systems help utilities and consumers manage electricity expenses more effectively. Moreover, energy storage enhances energy security by reducing dependence on imported fossil fuels and promoting local energy production and storage capabilities, thereby enhancing energy resilience and self-sufficiency.

## Environmental Impact

From an environmental perspective, energy storage supports sustainability goals by enabling greater penetration of renewable energy sources and reducing greenhouse gas emissions associated with fossil fuel-based electricity generation. By storing and dispatching renewable energy efficiently, energy storage systems help mitigate climate change impacts and improve air quality by reducing reliance on fossil fuel combustion.

In conclusion, energy storage technologies are pivotal for achieving a sustainable, reliable, and resilient energy system. By supporting grid stability, integrating renewable energy sources, optimizing electricity consumption, enhancing energy security, and mitigating environmental impacts, energy storage plays a critical role in shaping the future of energy. Continued advancements in energy storage technologies, coupled with supportive policies and market incentives, are essential for maximizing the benefits of energy storage and accelerating the transition towards a cleaner and more sustainable energy future.

## 6.2 Types of Energy Storage Systems (Batteries, Thermal, Mechanical)

Energy storage systems encompass a diverse range of technologies designed to store excess energy for later use, helping to balance supply and demand in electricity grids, enhance grid stability, and support the integration of renewable energy sources. This chapter explores the main types of energy storage systems: batteries, thermal storage, and mechanical storage.

### Batteries

Batteries are among the most versatile and widely deployed energy storage technologies, offering high efficiency, rapid response times, and scalability. They convert chemical energy into electrical energy during charging and vice versa during discharging.

## 1. Lithium-ion Batteries:

- **Description**: Lithium-ion batteries are the most common type used in energy storage applications due to their high energy density, efficiency, and long cycle life.
- **Applications**: They are employed in both grid-scale and distributed energy storage systems, ranging from small-scale residential batteries to large-scale utility storage projects.

## 2. Flow Batteries:

- **Description**: Flow batteries store energy in liquid electrolytes stored in external tanks, separated by a membrane.
- **Applications**: They offer advantages in terms of scalability and duration, making them suitable for grid-scale applications where longer-duration storage is required.

## Thermal Storage

Thermal energy storage systems store heat or cold for later use, utilizing materials that can store and release thermal energy based on temperature variations.

1. **Sensible Heat Storage:**

- **Description**: Sensible heat storage systems use materials such as water or rocks to store heat, which can be later used for heating applications or converted into electricity.
- **Applications**: They are used in concentrated solar power (CSP) plants to store heat collected from sunlight, allowing for continuous electricity generation even when the sun is not shining.

2. **Latent Heat Storage:**

- **Description**: Latent heat storage systems store and release energy during phase-change processes, such as melting and freezing of materials like phase-change materials (PCMs).
- **Applications**: They are utilized in refrigeration, air conditioning systems, and solar thermal energy storage to store excess heat for later use.

## Mechanical Storage

Mechanical energy storage systems store energy in the form of mechanical potential or kinetic energy, which is converted back into electricity when needed.

1. **Pumped Hydroelectric Storage:**

- **Description**: Pumped hydro storage systems store energy by pumping water uphill to a reservoir during periods of low electricity demand.
- **Applications**: They release stored water through turbines to generate electricity during peak demand periods, offering large-scale storage capacity and grid stability support.

## 2. Compressed Air Energy Storage (CAES):

- **Description**: CAES systems store energy by compressing air into underground caverns or storage tanks during periods of low demand.
- **Applications**: They release compressed air to drive turbines and generate electricity during peak demand periods, providing grid stability and supporting renewable energy integration.

## Applications and Benefits

Each type of energy storage system offers unique advantages and applications:

- **Grid Stability**: Energy storage systems improve grid stability by providing frequency regulation, voltage support, and grid balancing services.
- **Renewable Integration**: They enable the integration of intermittent renewable energy sources like solar and wind by storing excess energy and releasing it when renewable generation is low or when demand peaks.

- **Peak Shaving**: Energy storage systems reduce peak demand charges by storing electricity during off-peak hours and supplying it during peak demand periods, optimizing electricity consumption and reducing costs.

Energy storage systems are essential for enhancing the reliability, flexibility, and sustainability of modern energy systems. By diversifying energy storage technologies and optimizing their deployment in electricity grids, stakeholders can maximize the benefits of energy storage, support renewable energy integration, and promote a more resilient and efficient energy infrastructure. Continued advancements in energy storage technologies and supportive policies are crucial for accelerating the transition towards a cleaner and more sustainable energy future.

## 6.3 Selecting the Right Storage System for Your Home

Selecting the right energy storage system for your home involves considering various factors such as energy needs, budget, space availability, and compatibility with existing energy systems. Here's a comprehensive guide to help you choose the most suitable storage system:

**Assessing Your Energy Needs**

1. **Electricity Consumption**: Start by assessing your average daily electricity consumption. This will help determine the storage

capacity required to meet your household's energy needs during periods of low renewable energy generation or grid outages.

2. **Peak Demand**: Identify peak electricity demand periods in your home. Energy storage systems can help reduce peak demand charges by supplying stored energy during these times, potentially lowering your electricity bills.

**Types of Energy Storage Systems**

## 1. Battery Storage Systems:

- **Description**: Battery storage systems, such as lithium-ion batteries, are popular for residential applications due to their efficiency, scalability, and ability to store electricity for later use.
- **Applications**: They are suitable for storing excess energy from rooftop solar panels or grid electricity during off-peak hours, providing backup power during outages, and optimizing the self-consumption of solar energy.

## 2. Thermal Energy Storage:

- **Description**: Thermal storage systems store heat or cold for later use, typically using materials like water tanks or phase-change materials (PCMs) that absorb and release heat.
- **Applications**: They are ideal for homes with solar thermal systems, providing hot water storage or space heating, and reducing heating costs by storing excess heat generated during sunny periods.

### 3. Hybrid Systems:

- **Description**: Hybrid storage systems combine different storage technologies, such as batteries with solar panels or wind turbines, to maximize energy capture and storage efficiency.
- **Applications**: They offer flexibility in energy management, allowing homeowners to optimize renewable energy utilization and maintain energy resilience.

## Considerations for Selecting a Storage System

1. **System Size and Capacity**: Choose a storage system with adequate capacity to meet your household's energy needs during periods of high demand or low renewable energy generation. Consider factors like battery capacity (kWh), discharge rate, and compatibility with your energy consumption patterns.
2. **Installation and Space Requirements**: Evaluate the space available for installing the storage system. Batteries may require dedicated space indoors or outdoors, while thermal storage tanks need adequate room for installation and plumbing connections.
3. **Cost and Return on Investment (ROI)**: Compare upfront costs, installation expenses, and potential savings in electricity bills or peak demand charges over the system's lifespan. Calculate the ROI based on your energy consumption patterns and local utility rates to determine the economic feasibility of the investment.
4. **Maintenance and Warranty**: Consider the maintenance requirements and warranty coverage offered by storage system manufacturers. Ensure the system is reliable, with adequate support and service options available locally.

5. **Compatibility with Renewable Energy Sources**: If you have solar panels or plan to install them in the future, choose a storage system compatible with renewable energy sources. This ensures efficient energy capture, storage, and utilization, maximizing the benefits of renewable energy generation.

## Installation and Safety Considerations

1. **Professional Installation**: Hire qualified installers experienced in energy storage systems to ensure proper installation, and compliance with local building codes, and safety standards.
2. **Safety Features**: Ensure the storage system includes safety features such as overcharge protection, temperature regulation, and emergency shutdown mechanisms to prevent accidents and ensure safe operation.

Selecting the right energy storage system for your home involves assessing your energy needs, considering system types and applications, evaluating installation requirements and costs, and ensuring compatibility with renewable energy sources. By choosing a suitable storage system, homeowners can enhance energy efficiency, reduce electricity costs, and improve energy resilience, contributing to a sustainable and reliable energy future for their households.

## 6.4 Integration with Renewable Energy Sources

Integrating energy storage systems with renewable energy sources is pivotal for maximizing the efficiency, reliability, and sustainability of modern energy systems. This chapter explores the importance of integration and key considerations for effectively combining energy storage with renewable energy sources such as solar and wind power.

## Importance of Integration

Renewable energy sources like solar and wind power are inherently intermittent, fluctuating based on weather conditions and time of day. Energy storage systems play a crucial role in smoothing out these fluctuations by storing excess energy during periods of high generation and releasing it when renewable generation is low or when electricity demand peaks. This capability enhances grid stability, reduces reliance on fossil fuel-based backup power plants, and supports the reliable integration of renewable energy into the electricity grid.

## Key Considerations for Integration

### 1. Matching Supply and Demand:

Energy storage systems enable the alignment of renewable energy generation with electricity demand patterns. By storing surplus renewable energy during periods of high generation, storage systems ensure energy availability during peak demand periods or when renewable generation is limited.

## 2. Enhancing Energy Reliability:

Integrating energy storage with renewable energy sources enhances energy reliability by providing backup power during grid outages or periods of low renewable generation. Homeowners and businesses with renewable energy systems can maintain electricity supply continuity, reducing dependency on the main grid and enhancing energy resilience.

## 3. Optimizing Self-Consumption:

Energy storage systems optimize the self-consumption of solar energy generated by rooftop photovoltaic (PV) systems. Excess solar energy generated during the day can be stored in batteries for use during evenings or cloudy days, maximizing the utilization of renewable energy and reducing grid dependence.

## 4. Grid Support Services:

Energy storage systems provide grid support services such as frequency regulation, voltage control, and grid balancing. These services help stabilize the grid by mitigating fluctuations in renewable energy output and maintaining grid stability and reliability.

## 5. Economic Benefits:

Integration of energy storage with renewable energy sources offers economic benefits by reducing electricity costs through peak shaving and demand management strategies. The stored energy can be discharged during peak demand periods, minimizing peak electricity prices and optimizing energy consumption patterns.

### 6. Technological Compatibility:

Selecting energy storage systems that are compatible with renewable energy sources is essential for efficient energy capture, storage, and utilization. Battery storage systems, for example, are commonly integrated with solar PV systems to store excess solar energy for later use.

### 7. Policy and Regulatory Support:

Supportive policies and regulations, such as incentives for renewable energy and energy storage deployment, are crucial for promoting integration and accelerating the adoption of sustainable energy solutions. Clear regulatory frameworks encourage investment in renewable energy and storage technologies, driving innovation and market growth.

## Integration Techniques

### 1. Direct Coupling:

Direct coupling involves physically connecting energy storage systems with renewable energy sources, such as pairing batteries with solar PV systems. This approach maximizes energy capture and storage efficiency, directly utilizing renewable energy for onsite consumption or grid export.

## 2. Hybrid Systems:

Hybrid energy systems combine multiple renewable energy sources with complementary storage technologies, such as wind turbines with battery storage or solar PV with thermal energy storage. Hybrid systems optimize energy generation and storage capabilities, enhancing overall system efficiency and reliability.

## 3. Grid-Interactive Systems:

Grid-interactive systems enable bidirectional energy flow between energy storage systems, renewable energy sources, and the electricity grid. These systems can participate in grid services, providing ancillary services and supporting grid stability while optimizing renewable energy utilization.

Integrating energy storage systems with renewable energy sources is essential for advancing towards a sustainable and resilient energy future. By effectively combining storage technologies with solar, wind, and other renewable energy sources, stakeholders can enhance energy reliability, optimize energy consumption, reduce electricity costs, and support the transition to a low-carbon economy. Continued advancements in energy storage technologies, coupled with supportive policies and regulatory frameworks, will be key to maximizing the benefits of renewable energy integration and accelerating the global transition towards clean and sustainable energy systems.

# Chapter 7: Smart Home Technologies and Energy Management

Smart home technologies revolutionize energy management by integrating advanced digital solutions with household appliances and energy systems. This chapter explores the significance of smart home technologies in enhancing energy efficiency, optimizing consumption patterns, and improving overall sustainability.

**Importance of Smart Home Technologies**

Smart home technologies enable homeowners to monitor, control, and automate energy usage throughout their homes, resulting in several key benefits:

1. **Energy Efficiency**: Smart devices, such as smart thermostats, lighting systems, and appliances, optimize energy consumption by adjusting settings based on occupancy, time of day, and user preferences. Automated scheduling and energy-efficient modes reduce energy waste and lower utility bills.
2. **Real-time Monitoring**: Home energy management systems provide real-time insights into energy usage patterns, allowing homeowners to identify energy-intensive appliances or behaviors and take proactive measures to reduce energy consumption.
3. **Demand Response**: Smart home technologies support demand response programs by enabling appliances and devices to automatically adjust energy consumption during peak demand periods or in response to utility signals. This helps stabilize the grid and reduce strain on electricity networks.

**Key Components of Smart Home Technologies**

### 1. Smart Thermostats:

Functionality: Smart thermostats regulate home heating and cooling based on occupancy, preferences, and external temperature conditions. They learn user behavior to optimize energy use and can be controlled remotely via mobile apps.

### 2. Energy Monitoring Systems:

Capabilities: These systems track electricity, water, and gas consumption in real time, providing homeowners with detailed insights and analytics to promote energy conservation and efficiency.

### 3. Smart Lighting:

Features: Smart lighting systems use LED bulbs and sensors to adjust brightness and turn lights on/off automatically. They can be programmed to operate on schedules or in response to occupancy, reducing energy waste.

### 4. Energy-efficient Appliances:

Advantages: Smart appliances, such as refrigerators, washing machines, and dishwashers, feature energy-saving modes and connectivity for remote monitoring and control. They optimize energy use without compromising performance.

## Benefits of Smart Home Energy Management

1. **Cost Savings**: By optimizing energy consumption and reducing waste, smart home technologies help lower utility bills over time. Automated energy-saving measures and efficient appliance operation contribute to financial savings for homeowners.
2. **Environmental Impact**: Reduced energy consumption translates to lower carbon emissions and environmental impact. Smart home technologies support sustainability goals by promoting energy efficiency and responsible resource use.
3. **Enhanced Comfort and Convenience**: Automated climate control, lighting, and appliance management enhance home comfort and convenience. Remote access capabilities allow homeowners to adjust settings from anywhere, ensuring comfort while optimizing energy use.

## Integration with Renewable Energy Systems

1. **Solar PV Integration**: Smart home technologies can integrate with rooftop solar PV systems to optimize the self-consumption of solar energy. Homeowners can monitor solar production, store excess energy in batteries, and use smart appliances to maximize energy savings.
2. **Energy Storage Integration**: Smart energy management systems coordinate with energy storage solutions, such as batteries or

thermal storage, to store and dispatch energy based on real-time consumption patterns and utility rates.

**Future Trends and Innovations**

1. **Artificial Intelligence (AI):** AI-driven algorithms enhance energy management by predicting consumption patterns, optimizing energy use, and automating smart home functions based on user behavior and external factors.
2. **Internet of Things (IoT)**: The expansion of IoT devices and connectivity improves interoperability and data exchange between smart home systems, appliances, and utilities, facilitating seamless energy management and grid interaction.

Smart home technologies and energy management systems empower homeowners to achieve energy efficiency, reduce costs, and contribute to sustainability efforts. By integrating smart devices, monitoring systems, and energy-efficient appliances, households can optimize energy consumption, enhance comfort, and leverage renewable energy resources effectively. Continued innovation and adoption of smart home technologies are essential for advancing towards a more resilient, efficient, and sustainable future in residential energy management.

## 7.1 Overview of Smart Home Energy Systems

Smart home energy systems represent a transformative approach to managing household energy consumption through interconnected devices, advanced sensors, and automated controls. This section provides an overview of smart home energy systems, highlighting their components, benefits, and integration capabilities.

**Components of Smart Home Energy Systems**

**1. Smart Thermostats:**

Functionality: Smart thermostats regulate indoor temperatures based on user preferences, occupancy, and external weather conditions. They learn household habits to optimize heating and cooling efficiency and can be controlled remotely via smartphone apps.

**2. Energy Monitoring Systems:**

Capabilities: Energy monitoring systems track electricity, water, and gas consumption in real-time. They provide homeowners with detailed insights into energy usage patterns, enabling informed decisions to reduce waste and improve efficiency.

**3. Smart Lighting:**

Features: Smart lighting systems use energy-efficient LED bulbs and sensors to adjust lighting levels automatically. They can be programmed to turn lights on/off based on occupancy, time of day, or ambient light levels, contributing to energy savings.

### 4. Energy-efficient Appliances:

Advantages: Smart appliances, such as refrigerators, washing machines, and dishwashers, incorporate energy-saving features and connectivity for remote monitoring and control. They optimize energy consumption without compromising performance.

**Benefits of Smart Home Energy Systems**

### 1. Energy Efficiency:

Smart home energy systems optimize energy use by adjusting device settings and schedules based on real-time data and user preferences. Automated energy-saving measures reduce consumption and lower utility bills over time.

### 2. Cost Savings:

By monitoring and controlling energy consumption, smart home technologies help homeowners save on utility expenses. Energy-efficient practices and automated adjustments minimize waste, contributing to long-term financial savings.

### 3. Environmental Impact:

Reduced energy consumption through smart home systems leads to lower carbon emissions and environmental impact. Energy-efficient appliances and automated controls promote sustainable living practices and support environmental conservation efforts.

### 4. Enhanced Comfort and Convenience:

Automated climate control, lighting adjustments, and appliance management enhance home comfort and convenience. Remote access features allow homeowners to monitor and adjust settings from anywhere, ensuring optimal comfort while maximizing energy efficiency.

## Integration Capabilities

### 1. Renewable Energy Integration:

Smart home energy systems can integrate with renewable energy sources like solar photovoltaic (PV) systems. They optimize the self-consumption of solar energy by coordinating energy storage, appliance operation, and home energy use patterns.

### 2. Energy Storage Systems:

Integration with energy storage solutions, such as batteries or thermal storage, allows smart home systems to store excess energy for later use. They optimize energy dispatch based on real-time consumption data and utility pricing signals.

### 3. Grid Interaction and Demand Response:

Smart home energy systems support grid interaction through demand response programs. They can automatically adjust energy consumption during peak demand periods or in response to utility signals, contributing to grid stability and efficiency.

**Future Trends and Innovations**

### 1. Artificial Intelligence (AI) and Machine Learning:

AI-powered algorithms enhance energy management by predicting consumption patterns, optimizing device operations, and learning user behaviors to automate smart home functions effectively.

### 2. Internet of Things (IoT) Expansion:

Continued growth in IoT devices and connectivity improves interoperability and data exchange among smart home systems, appliances, and utility networks. Enhanced IoT integration supports seamless energy management and grid interaction capabilities.

Smart home energy systems represent a significant advancement in residential energy management, offering homeowners enhanced control, efficiency, and sustainability. By integrating smart devices, monitoring systems, and energy-efficient appliances, households can optimize energy consumption, reduce costs, and support environmental goals. Continued innovation in smart home technologies, coupled with widespread adoption and supportive policies, will play a crucial role in advancing towards a more efficient and sustainable future in residential energy use.

## 7.2 Smart Grids and Their Role in Energy Efficiency

Smart grids represent a modernized electrical grid infrastructure that leverages digital technology to enhance reliability, efficiency, and sustainability in energy distribution and consumption. This section explores the concept of smart grids, their components, benefits, and their pivotal role in improving energy efficiency.

**Understanding Smart Grids**

### 1. Definition and Components:

Definition: A smart grid integrates advanced communication, control, and monitoring technologies into the traditional electricity grid infrastructure.

Components: Key components include smart meters, sensors, automation systems, and digital communication networks that enable real-time data exchange and grid management.

### 2. Real-time Monitoring and Control:

Smart grids enable utilities to monitor electricity supply and demand in real time. Sensors and meters collect data on energy usage patterns, grid conditions, and equipment performance, allowing utilities to optimize grid operations and respond quickly to disruptions.

### 3. Two-way Communication:

Smart grids facilitate two-way communication between utilities and consumers. Smart meters and digital sensors provide consumers with detailed information about their energy usage and enable utilities to remotely manage electricity flow, implement demand response programs, and optimize energy distribution.

**Benefits of Smart Grids**

### 1. Enhanced Grid Reliability and Resilience:

Smart grids improve grid reliability by detecting and responding to power outages more quickly. Automated systems reroute power around affected areas, minimizing downtime and enhancing overall grid resilience against disruptions.

### 2. Energy Efficiency and Optimization:

By monitoring real-time data on energy consumption and grid conditions, smart grids optimize energy distribution. Utilities can reduce energy losses during transmission and distribution, improve voltage management, and prioritize renewable energy integration.

### 3. Demand Response and Peak Shaving:

Smart grids support demand response initiatives by enabling utilities to adjust electricity usage during peak demand periods. Time-of-use pricing and smart metering encourage consumers to shift energy-intensive activities to off-peak hours, reducing strain on the grid and lowering electricity costs.

### 4. Integration of Renewable Energy Sources:

Smart grids facilitate the integration of renewable energy sources like solar and wind power. Advanced forecasting tools and grid management systems optimize renewable energy generation, storage, and distribution, ensuring reliable and efficient grid operation.

## Role of Smart Grids in Energy Efficiency

### 1. Energy Conservation and Management:

Smart grids promote energy conservation through improved energy management practices. Consumers gain insights into their energy usage patterns and can adjust behaviors or upgrade to energy-efficient appliances based on real-time data and recommendations.

## 2. Grid Modernization and Infrastructure Investment:

Smart grid investments modernize aging grid infrastructure, enhancing its capacity, reliability, and efficiency. Upgraded transmission and distribution networks accommodate growing electricity demand and support the adoption of new technologies and energy solutions.

## 3. Environmental Sustainability:

By optimizing energy distribution and reducing energy losses, smart grids contribute to environmental sustainability. Increased renewable energy integration and reduced reliance on fossil fuels lower greenhouse gas emissions, supporting global climate goals.

## Future Directions and Innovations

### 1. Advanced Metering Infrastructure (AMI):

AMI systems enhance smart grid capabilities by providing real-time data on energy consumption and grid performance. Integration with smart meters and digital communication networks improves grid visibility and operational efficiency.

### 2. Distributed Energy Resources (DERs):

Smart grids support the integration of distributed energy resources such as rooftop solar panels, energy storage systems, and electric vehicles.

Coordination of DERs enhances grid flexibility, reliability, and resilience while empowering consumers to participate in energy markets.

Smart grids play a crucial role in transforming traditional electricity grids into modern, efficient, and sustainable systems. By leveraging digital technologies, real-time data analytics, and advanced communication networks, smart grids enhance grid reliability, optimize energy distribution, support renewable energy integration, and promote energy efficiency. Continued investment in smart grid infrastructure and regulatory support will accelerate the transition towards a cleaner, more resilient energy future while meeting growing electricity demands and addressing environmental challenges.

# 7.3 Home Automation for Energy Conservation

Home automation systems empower homeowners to manage and optimize energy consumption through automated controls, smart devices, and real-time data insights. This section explores the concept of home automation for energy conservation, highlighting its benefits, components, and practical applications.

## Understanding Home Automation

### 1. Definition and Components:

Definition: Home automation refers to the use of interconnected devices and systems to automate household tasks and operations, including energy management.

Components: Key components include smart thermostats, lighting controls, appliances, sensors, and central control hubs that enable remote monitoring and control of home systems.

## 2. Automated Energy Management:

Home automation systems monitor energy usage in real time and automate energy-saving behaviors based on predefined settings, occupancy patterns, and external factors like weather conditions. Automated controls adjust device operation, lighting, heating, and cooling to optimize energy efficiency.

## 3. Integration with Smart Devices:

Integration with smart devices enables seamless communication and interoperability between different systems within the home. Smart thermostats, lighting controls, and appliances can be synchronized to operate efficiently and respond to user preferences and environmental conditions.

**Benefits of Home Automation for Energy Conservation**

## 1. Energy Efficiency:

Automated controls and scheduling optimize energy consumption by adjusting heating, cooling, and lighting based on occupancy and time of day. Energy-efficient modes and automated shutdowns reduce standby power consumption, lowering overall energy usage and utility bills.

## 2. Real-time Monitoring and Insights:

Home automation systems provide real-time data on energy consumption patterns, appliance usage, and home occupancy. Detailed insights empower homeowners to identify energy-intensive behaviors or devices and make informed decisions to improve efficiency.

## 3. Enhanced Convenience and Comfort:

Automated climate control, lighting adjustments, and appliance management enhance home comfort and convenience. Remote access via smartphone apps allows homeowners to monitor and adjust settings from anywhere, ensuring optimal comfort while minimizing energy waste.

## 4. Demand Response Participation:

Home automation systems support participation in demand response programs by automatically reducing energy consumption during peak demand periods or in response to utility signals. Smart devices adjust operation to align with grid needs, contributing to grid stability and reliability.

**Practical Applications of Home Automation**

### 1. Smart Thermostats:

Smart thermostats regulate home heating and cooling based on occupancy and user preferences. Learning algorithms adapt to household schedules and environmental conditions to optimize energy use and reduce heating and cooling costs.

### 2. Lighting Controls:

Smart lighting systems use energy-efficient LED bulbs and motion sensors to adjust lighting levels based on occupancy and natural light levels. Automated schedules and remote controls minimize unnecessary lighting and reduce electricity consumption.

### 3. Appliance Automation:

Smart appliances, such as refrigerators, washing machines, and dishwashers, incorporate energy-saving features and programmable settings. Automation schedules optimize appliance operation for energy efficiency without compromising performance.

**Future Trends and Innovations**

### 1. Integration with AI and Machine Learning:

AI-driven algorithms enhance home automation capabilities by predicting energy usage patterns, learning user behaviors, and optimizing device operations for maximum efficiency and comfort.

## 2. Expansion of IoT Devices:

Continued growth in IoT devices and connectivity improves interoperability and data exchange among smart home systems. Enhanced integration supports advanced energy management strategies and seamless interaction with utility services.

Home automation systems offer homeowners a powerful tool to conserve energy, reduce costs, and enhance comfort through automated controls and smart devices. By optimizing energy consumption, providing real-time insights, and supporting demand response initiatives, home automation contributes to sustainable living practices and promotes environmental stewardship. Continued innovation in technology and widespread adoption of home automation solutions will drive further improvements in energy efficiency, comfort, and convenience, paving the way for a smarter and more sustainable future in residential energy management.

# 7.4 Monitoring and Managing Energy Use

Effective monitoring and management of energy use are essential for optimizing efficiency, reducing costs, and promoting sustainability in residential settings. This chapter explores various methods, technologies, and strategies for monitoring and managing energy consumption without bullet points.

# Understanding Energy Monitoring

Monitoring energy use involves tracking and analyzing electricity consumption patterns, identifying inefficiencies, and implementing measures to reduce waste. Modern energy monitoring systems utilize advanced sensors, smart meters, and digital platforms to provide real-time data insights into household energy use.

## Components of Energy Monitoring Systems

### Smart Meters:

Smart meters measure electricity consumption at regular intervals and transmit data to utility companies or homeowners. They provide detailed information on energy usage, peak demand periods, and billing accuracy, enabling informed energy management decisions.

### Energy Monitoring Devices:

Energy monitoring devices connect to electrical panels or individual appliances to track energy consumption in real time. They monitor power usage, voltage levels, and energy costs, offering homeowners visibility into appliance efficiency and overall energy performance.

**Data Analytics Platforms:**

Data analytics platforms analyze energy data collected from smart meters and monitoring devices. They generate insights on consumption trends, peak demand times, and potential energy-saving opportunities. Visualization tools and reports help homeowners understand usage patterns and make informed decisions.

**Benefits of Energy Monitoring and Management**

**Optimized Energy Consumption:**

Real-time monitoring identifies energy-intensive appliances and behaviors, enabling homeowners to adjust settings and schedules for maximum efficiency. By understanding usage patterns, households can reduce standby power, optimize heating and cooling, and minimize wasted energy.

**Cost Savings:**

Monitoring energy use allows homeowners to identify energy-saving opportunities and reduce utility bills. By implementing efficiency measures and adjusting consumption habits, households can lower electricity costs over time and achieve financial savings.

## Environmental Impact:

Effective energy management reduces carbon emissions and environmental impact associated with energy production. By conserving energy and promoting sustainable practices, homeowners contribute to environmental stewardship and support climate goals.

## Strategies for Managing Energy Use

### Set Energy Goals:

Establish energy-saving goals based on household priorities, such as reducing overall consumption, optimizing peak demand usage, or integrating renewable energy sources.

### Implement Efficiency Measures:

Upgrade to energy-efficient appliances, install programmable thermostats, and seal drafts to minimize energy losses. Smart home automation and scheduling optimize device operation for maximum efficiency.

### Monitor and Adjust:

Regularly monitor energy consumption data and adjust habits or settings accordingly. Analyze trends, review energy bills, and use monitoring tools to track progress toward efficiency goals.

**Technology Integration and Innovations**

**Integration with Smart Home Systems:**

Energy monitoring integrates with smart home automation systems to automate energy-saving actions. Smart devices adjust settings based on occupancy, weather conditions, and time of day, optimizing energy use without manual intervention.

**Blockchain and Energy Trading:**

Blockchain technology enables peer-to-peer energy trading and transparent energy transactions. It supports decentralized energy management, empowers consumers to buy and sell excess renewable energy, and promotes energy resilience.

**Future Directions**

Continuous advancements in energy monitoring technology, data analytics, and smart grid integration will enhance capabilities for managing energy use effectively. By leveraging real-time data insights, automation, and sustainable practices, homeowners can achieve greater energy efficiency, reduce costs, and contribute to a cleaner, more resilient energy future.

Monitoring and managing energy use are critical aspects of sustainable living and cost-effective energy management. By adopting energy monitoring systems, implementing efficiency measures, and leveraging technology innovations, homeowners can optimize energy consumption, reduce environmental impact, and achieve long-term savings. Embracing proactive energy management practices supports global sustainability efforts and fosters a smarter, more resilient energy ecosystem for future generations.

# Chapter 8: Financing and Incentives for Green Energy Projects

Financing and incentives play a crucial role in facilitating the adoption of green energy projects, enabling homeowners and businesses to invest in renewable energy technologies and sustainable practices. This chapter explores various financial mechanisms, incentives, and funding options available to support green energy initiatives.

**Understanding Financing Options**

## 1. Loans and Mortgages:

Home Improvement Loans: Many financial institutions offer specialized loans for home improvements, including energy-efficient upgrades such as solar panel installations or energy-efficient appliances. These loans often feature competitive interest rates and flexible repayment terms tailored to energy project budgets.

Energy Efficient Mortgages (EEMs): EEMs allow homeowners to finance energy-saving upgrades as part of their mortgage loan. Lenders assess the potential energy savings when determining loan eligibility, offering favorable terms and potentially higher loan amounts to cover upfront costs.

## 2. Leasing and Power Purchase Agreements (PPAs):

Solar Leasing: In solar leasing arrangements, homeowners or businesses lease solar panels from a provider who installs and maintains the system. Monthly lease payments replace upfront costs, making solar energy more accessible without the initial financial burden of ownership.

Power Purchase Agreements (PPAs): PPAs allow homeowners or businesses to purchase solar electricity at a predetermined rate from a solar provider. The provider installs and maintains the solar panels, and customers benefit from predictable electricity costs typically lower than utility rates.

## Incentives and Rebates

## 1. Federal Tax Credits:

Solar Investment Tax Credit (ITC): The ITC offers a federal tax credit of up to 26% of the cost of solar energy systems installed on residential or commercial properties. The credit reduces homeowners' tax liability, making solar energy more affordable and attractive.

Renewable Energy Production Incentives: Some states or utility companies offer production incentives for renewable energy generation, such as feed-in tariffs or performance-based incentives for solar or wind installations.

## 2. State and Local Incentives:

Rebates and Grants: Many states and local governments provide rebates or grants for energy-efficient upgrades and renewable energy installations. These incentives offset upfront costs and encourage investment in green energy technologies.

Property Tax Exemptions: Property tax exemptions or deductions may apply to homes with renewable energy systems, reducing overall property tax burdens for homeowners who invest in solar, wind, or geothermal energy.

## Financing Mechanisms for Businesses

## 1. Commercial Property Assessed Clean Energy (C-PACE):

C-PACE programs allow commercial property owners to finance energy efficiency and renewable energy projects through property tax assessments. Repayments are made over time through property tax bills, leveraging long-term energy cost savings to finance upfront investments.

## 2. Energy Efficiency Loans and Grants:

Businesses may qualify for energy efficiency loans or grants through government programs, financial institutions, or utility-sponsored initiatives. These funds support energy audits, efficiency upgrades, and renewable energy installations, enhancing operational sustainability and cost savings.

**Community and Cooperative Initiatives**

### 1. Community Solar Programs:

Community solar projects enable multiple participants to invest in a shared solar installation located off-site. Participants receive credits on their electricity bills for the energy produced by their share of the solar array, providing access to solar benefits for renters or homeowners with shaded roofs.

### 2. Cooperative Purchasing:

Cooperative purchasing allows groups of homeowners or businesses to leverage collective purchasing power for bulk discounts on solar installations or energy-efficient products. Cooperative models lower individual costs and streamline the adoption of green energy solutions within communities.

Financing and incentives are instrumental in overcoming financial barriers and accelerating the adoption of green energy projects. By leveraging loans, leases, tax credits, and local incentives, homeowners and businesses can invest in renewable energy systems, energy-efficient upgrades, and sustainable practices to reduce energy costs, enhance property value, and contribute to environmental conservation. Continued support from governments, utilities, and financial institutions is essential to expanding access to affordable green energy solutions and advancing towards a cleaner, more resilient energy future.

# 8.1 Understanding the Costs and Savings

Understanding the costs and savings associated with green energy projects is essential for homeowners and businesses considering investments in renewable energy systems and energy-efficient upgrades. This section explores the financial aspects of green energy, including upfront costs, long-term savings, and factors influencing economic feasibility.

**Upfront Costs of Green Energy Projects**

Investing in green energy technologies typically involves significant upfront costs that vary based on the type and scale of the project:

1. **Solar Photovoltaic (PV) Systems:**

- The cost of solar PV systems includes equipment (solar panels, inverters, mounting hardware), installation labor, and permits. Costs can vary widely depending on system size, location, roof complexity, and local labor rates.
- Average residential solar PV system costs have decreased over the years due to technological advancements and economies of scale but can still range from several thousand to tens of thousands of dollars.

## 2. Wind Turbines:

- Small-scale wind turbines for residential use require substantial upfront investment for turbine purchase, installation, and associated infrastructure (tower, foundation, electrical connections).
- Costs depend on turbine size, wind resource availability, permitting requirements, and installation complexity. Larger turbines generally have higher upfront costs but may offer greater energy production and savings over time.

## 3. Geothermal Heat Pumps:

- Geothermal heat pump systems involve upfront costs for equipment (heat pump unit, ground loop or well system), installation labor, and drilling or excavation if required.
- Costs vary based on system size, soil conditions, property size, and local geology. Initial investments can be significant but are offset by long-term energy savings and reduced operational costs.

## 4. Energy Efficiency Upgrades:

- Energy-efficient upgrades such as insulation, windows, HVAC systems, and lighting fixtures require upfront investment but offer immediate and ongoing savings through reduced energy consumption.
- Costs depend on the scope of upgrades, building size, existing infrastructure, and regional energy prices. Return on investment

(ROI) calculations consider energy savings over the upgrade's lifespan.

**Long-term Savings and Return on Investment (ROI)**

Despite initial investment requirements, green energy projects can yield substantial long-term savings and financial benefits:

1. **Energy Savings:**

- Renewable energy systems generate electricity or provide heating/cooling at lower costs than traditional utility rates. Solar PV systems produce electricity from sunlight, reducing or eliminating electricity bills depending on system size and local solar conditions.
- Geothermal heat pumps use ground or water source heat to heat and cool homes efficiently, often achieving significant energy savings compared to conventional HVAC systems.

2. **Return on Investment (ROI):**

- ROI for green energy projects considers upfront costs, ongoing savings from reduced energy bills, incentives, and tax credits. Solar PV systems typically have shorter payback periods due to favorable incentives like the federal Investment Tax Credit (ITC).
- Geothermal heat pumps and energy efficiency upgrades also offer attractive ROIs, with savings accumulating over the system's

lifespan. ROI calculations factor in energy price trends, system longevity, maintenance costs, and financing terms.

### 3. Increased Property Value:

- Green energy features enhance property value and marketability. Homes with solar PV systems, geothermal heat pumps, or high-efficiency ratings often command higher sale prices and attract environmentally-conscious buyers seeking reduced operating costs.

## Factors Influencing Economic Feasibility

Several factors influence the economic feasibility of green energy projects:

### 1. Local Incentives and Rebates:

Government incentives, tax credits, rebates, and grants reduce upfront costs and improve project economics. Incentive availability varies by location and may influence project timing and affordability.

### 2. Energy Prices and Utility Rates:

Higher energy prices increase the financial attractiveness of renewable energy and efficiency projects. Local utility rates impact potential savings from reduced energy consumption or grid-connected renewable energy generation.

### 3. Financing Options:

Access to low-interest loans, leasing arrangements, or energy performance contracts can lower upfront financial barriers and improve cash flow for green energy investments. Favorable financing terms enhance project affordability and ROI.

Understanding the costs and savings associated with green energy projects is critical for making informed decisions about sustainable investments. While upfront costs can be substantial, long-term savings, reduced energy bills, and potential property value increases justify the initial investment. Government incentives, favorable financing options, and local energy economics play pivotal roles in enhancing project affordability and accelerating the adoption of green energy solutions. By evaluating upfront costs, calculating potential savings, and considering economic feasibility factors, homeowners and businesses can embrace renewable energy technologies and energy-efficient practices to achieve financial benefits and contribute to environmental sustainability.

## 8.2 Government Incentives and Rebates

Government incentives and rebates play a pivotal role in promoting the adoption of green energy technologies and energy-efficient practices. This chapter explores various incentives, tax credits, grants, and rebate programs offered by governments at federal, state, and local levels to support renewable energy projects and encourage sustainability efforts.

# Federal Government Incentives

### 1. Investment Tax Credit (ITC):

The Federal Investment Tax Credit (ITC) is a significant financial incentive for residential and commercial solar energy systems. It allows eligible taxpayers to deduct a percentage of the cost of installing a solar energy system from their federal taxes.

As of 2024, the ITC offers a tax credit of 26% for solar systems installed by the end of 2022. The credit percentage decreases to 22% for systems installed in 2023 and beyond, providing substantial upfront cost savings for solar investments.

### 2. Production Tax Credit (PTC) and Investment Tax Credit (ITC) for Wind Energy:

The Production Tax Credit (PTC) and Investment Tax Credit (ITC) are incentives for wind energy projects. The PTC provides a per-kilowatt-hour tax credit for electricity generated from qualified wind facilities, while the ITC offers a credit based on capital investment in wind energy installations.

### 3. Residential Renewable Energy Tax Credit:

Homeowners may qualify for a Residential Renewable Energy Tax Credit for installing eligible renewable energy systems, such as solar

PV, wind, geothermal heat pumps, and fuel cell systems. The credit covers a percentage of installation costs, helping offset upfront expenses.

**State and Local Government Incentives**

**1. State Solar and Renewable Energy Rebates:**

Many states offer cash rebates or performance-based incentives for residential and commercial solar energy installations. Rebate amounts vary by state and utility company, providing financial incentives based on system size, energy production, and efficiency ratings.

**2. Property Tax Exemptions and Assessments:**

Some states and local jurisdictions provide property tax exemptions, assessments, or deductions for homes and businesses with renewable energy systems or energy-efficient upgrades. These incentives reduce property tax burdens, enhancing project economics and affordability.

**3. Sales Tax Exemptions:**

Sales tax exemptions or reductions may apply to the purchase and installation of renewable energy systems, making clean energy technologies more cost-effective for homeowners and businesses. Eligibility criteria and exemption amounts vary by state.

**Utility-Sponsored Incentive Programs**

### 1. Rebates and Performance-Based Incentives:

Utility companies administer rebate programs and performance-based incentives to promote energy efficiency upgrades and renewable energy installations. Rebates may cover a portion of project costs, encouraging customers to invest in energy-saving measures.

### 2. Net Metering and Feed-in Tariffs:

Net metering allows homeowners and businesses with renewable energy systems to offset their electricity bills by exporting surplus energy to the grid. Utilities credit customers for excess energy production at retail rates, enhancing the financial return on renewable energy investments.

Feed-in tariffs provide fixed payments or premiums for renewable energy generated and fed into the grid, offering long-term revenue streams and financial predictability for renewable energy producers.

**Grants and Financing Programs**

### 1. Energy Efficiency Grants:

Federal, state, and local governments offer grants to support energy efficiency projects, including insulation upgrades, energy-efficient appliances, and building retrofits. Grants provide non-repayable funding to offset project costs and promote sustainable building practices.

2. **Low-Interest Loans and Financing:**

Government-sponsored or subsidized loans provide low-interest financing options for renewable energy projects and energy efficiency improvements. Financing programs reduce upfront financial barriers, facilitate project implementation, and support long-term savings through reduced energy costs.

Government incentives and rebates significantly reduce the upfront costs of green energy projects, improve project economics, and accelerate the adoption of renewable energy technologies. By leveraging federal tax credits, state rebates, utility incentives, and financing programs, homeowners and businesses can invest in solar, wind, geothermal, and energy efficiency projects with enhanced affordability and financial return. Continued support and expansion of incentive programs are essential to promoting sustainability, reducing greenhouse gas emissions, and advancing toward a cleaner, more resilient energy future. Understanding available incentives and consulting with local authorities or energy experts can help stakeholders maximize financial benefits and achieve environmental goals through strategic green energy investments.

## 8.3 Financing Options and Strategies

Financing options and strategies are critical considerations for homeowners and businesses seeking to invest in green energy projects and energy-efficient upgrades. This chapter explores various financing mechanisms, strategies, and considerations to facilitate affordable and accessible funding for renewable energy installations and sustainability initiatives.

**Understanding Financing Options**

### 1. Solar Loans:

Solar loans are specialized financing products designed for residential and commercial solar installations. These loans offer fixed or variable interest rates, and flexible repayment terms, and may be secured or unsecured.

Benefits include lower upfront costs compared to outright purchases, enabling homeowners and businesses to spread payments over time while enjoying immediate energy savings from solar energy production.

### 2. Energy Efficiency Loans:

Energy efficiency loans finance energy-saving upgrades such as insulation, HVAC systems, lighting retrofits, and building envelope improvements. Loans may be available through financial institutions, government programs, or utility-sponsored initiatives.

Terms typically align loan repayments with energy cost savings, ensuring that monthly utility bill reductions offset loan payments, resulting in net-positive cash flow for borrowers.

### 3. Property-Assessed Clean Energy (PACE) Financing:

PACE financing allows property owners to finance renewable energy installations and energy efficiency upgrades through assessments on property tax bills. Repayments are tied to property taxes and may transfer to new property owners upon sale.

PACE programs offer long repayment terms, and competitive interest rates, and may cover 100% of project costs. Eligibility and program availability vary by location, with some jurisdictions supporting residential, commercial, and industrial projects.

## Leasing and Power Purchase Agreements (PPAs)

### 1. Solar Leases:

Solar leases enable homeowners and businesses to "rent" solar panels from a third-party provider. The provider installs, owns, and maintains the system, while the customer pays a fixed monthly lease payment for solar electricity generated.

Leasing arrangements eliminate upfront costs and maintenance responsibilities, making solar energy accessible to customers without the capital investment required for ownership.

### 2. Power Purchase Agreements (PPAs):

PPAs allow homeowners and businesses to purchase solar electricity from a third-party provider at a predetermined rate per kilowatt-hour.

Providers install, operate, and maintain the solar system, and customers benefit from predictable electricity costs lower than utility rates.

PPAs typically include long-term contracts with fixed or escalating rates, providing budget certainty and shielding customers from electricity price volatility.

**Government and Utility-Sponsored Programs**

**1. Grants and Rebates:**

Government grants and rebates provide non-repayable funding to offset upfront costs of renewable energy projects and energy efficiency upgrades. Programs vary by jurisdiction and may target specific technologies or sustainability goals.

Rebate amounts and eligibility criteria depend on project scope, energy savings potential, and program guidelines established by federal, state, or local authorities.

**2. Utility Incentives:**

Utility companies administer incentive programs to encourage customers to invest in energy efficiency measures and renewable energy installations. Incentives may include cash rebates, performance-based incentives, and net metering arrangements.

Utility incentives reduce payback periods and improve financial returns on green energy investments, supporting customer adoption of sustainable technologies and grid stability objectives.

**Financing Strategies for Businesses**

## 1. Commercial Property Assessed Clean Energy (C-PACE):

C-PACE financing enables commercial property owners to finance energy efficiency upgrades and renewable energy projects through property tax assessments. Repayments are made over time through property tax bills, leveraging energy cost savings to finance investments.

C-PACE programs support large-scale projects, offer competitive financing terms, and align repayments with property improvements that enhance building value and operational efficiency.

## 2. Energy Performance Contracts (EPCs):

Energy service companies (ESCOs) offer EPCs to finance and implement energy efficiency projects for businesses. ESCOs guarantee energy savings through upgraded equipment and systems, with contract terms based on performance metrics.

EPCs allow businesses to implement energy upgrades with minimal upfront costs, relying on projected energy savings to cover financing and operational expenses over the contract term.

Effective financing options and strategies are essential for overcoming financial barriers and accelerating the adoption of green energy projects and energy efficiency measures. By leveraging solar loans, energy efficiency financing, PACE programs, leasing agreements, and government incentives, homeowners and businesses can make sustainable investments with improved affordability and financial return. Collaboration with financial institutions, energy providers, and government agencies facilitates access to favorable terms, reduces project risks, and promotes environmental stewardship through strategic green energy financing. Continued innovation and expansion of financing mechanisms are crucial to advancing clean energy adoption, achieving energy savings, and mitigating climate impacts for a sustainable future.

## 8.4 Long-term Financial Benefits of Green Energy Investments

Investing in green energy technologies and energy efficiency measures offers significant long-term financial benefits for homeowners and businesses. This chapter explores the economic advantages, cost savings, and financial returns associated with sustainable energy investments over time.

**Reduced Energy Costs**

1. **Energy Savings:**

Green energy investments, such as solar PV systems, wind turbines, geothermal heat pumps, and energy-efficient upgrades, reduce reliance on grid electricity and fossil fuels. These technologies generate or conserve energy cost-effectively, leading to lower monthly utility bills.

Solar PV systems convert sunlight into electricity, offsetting or eliminating electricity purchases from utilities. Geothermal heat pumps harness stable ground temperatures for heating and cooling, reducing HVAC energy consumption year-round.

## 2. Fixed or Predictable Energy Costs:

Renewable energy systems, such as solar leases or power purchase agreements (PPAs), provide fixed or predictable electricity rates over extended contract terms. This shields homeowners and businesses from fluctuating energy prices and inflationary pressures associated with fossil fuels.

## Return on Investment (ROI)

### 1. Payback Periods:

Green energy investments typically have finite payback periods, during which upfront costs are recouped through energy savings or revenue generation. Solar PV systems, for example, often achieve payback within 5 to 10 years depending on system size, location, and incentive availability.

Geothermal heat pumps and energy efficiency upgrades offer shorter payback periods with immediate energy cost reductions and ongoing financial benefits throughout their operational lifetimes.

### 2. Increased Property Value:

Properties equipped with renewable energy systems or high-efficiency features generally command higher resale values and market premiums. Homebuyers prioritize energy-efficient homes for reduced operating costs, environmental benefits, and enhanced comfort.

**Financial Incentives and Tax Benefits**

### 1. Federal Tax Credits:

The Federal Investment Tax Credit (ITC) offers homeowners and businesses a percentage-based tax credit for installing solar PV, wind, and geothermal systems. Tax credits reduce income tax liability, enhancing project affordability and financial returns.

Other federal incentives include production tax credits (PTCs) for wind energy and residential energy efficiency tax credits for qualified home improvements.

### 2. State and Local Incentives:

State-specific rebates, grants, property tax incentives, and sales tax exemptions further reduce upfront costs and improve project economics. Incentives vary by location and support renewable energy deployment, energy independence, and climate resilience initiatives.

**Environmental and Social Benefits**

### 1. Carbon Footprint Reduction:

Green energy investments contribute to environmental sustainability by reducing greenhouse gas emissions and reliance on finite fossil fuel resources. Solar, wind, and geothermal technologies offer clean, renewable alternatives to traditional energy sources.

Environmental stewardship aligns with corporate social responsibility (CSR) goals, enhances brand reputation, and attracts environmentally conscious consumers and stakeholders.

## 2. Energy Independence and Resilience:

On-site renewable energy generation enhances energy independence and resilience against utility disruptions, extreme weather events, and grid failures. Battery storage systems and microgrid solutions further enhance energy security and reliability.

## Economic Growth and Job Creation

## 1. Economic Stimulus:

Investments in green energy infrastructure and clean technologies stimulate economic growth, create local job opportunities, and foster innovation across renewable energy sectors. The renewable energy industry supports diverse employment in manufacturing, installation, maintenance, and research.

## 2. Long-term Economic Stability:

Transitioning to sustainable energy sources mitigates energy price volatility, reduces trade deficits associated with fossil fuel imports, and promotes long-term economic stability. Green energy investments diversify energy portfolios and support sustainable economic development goals.

Green energy investments deliver compelling long-term financial benefits, including reduced energy costs, enhanced property value, favorable ROI, and access to financial incentives and tax benefits. By embracing renewable energy technologies and energy efficiency measures, homeowners and businesses achieve financial savings, environmental sustainability, and resilience against energy market fluctuations. Continued policy support, innovation in clean energy technologies, and public-private partnerships are essential to maximizing economic opportunities, advancing energy transition goals, and securing a sustainable future for generations to come.

# Conclusion:

In "Green Energy Solutions: A Guide to Practical Strategies for Implementing Renewable Energy Technologies in Your Home and Energy Storage," we have explored the transformative potential of renewable energy and energy storage solutions. This journey through sustainable practices and technological innovations has illuminated practical strategies for homeowners and businesses alike to adopt green energy technologies and enhance energy efficiency.

## Embracing Sustainability

The adoption of renewable energy technologies, such as solar photovoltaics, wind turbines, geothermal heat pumps, and biomass systems, offers more than just a means to reduce utility bills. It represents a commitment to environmental stewardship, reducing carbon footprints, and mitigating climate change impacts. Each chapter has underscored the role of green energy in fostering a cleaner, more resilient energy future for communities worldwide.

## Practical Implementation Strategies

From understanding the basics of renewable energy to navigating financing options and leveraging government incentives, this guide has equipped readers with the knowledge to make informed decisions. Whether through solar leases, energy efficiency loans, or participation in community solar programs, the path to sustainable energy adoption is clearer and more accessible than ever before.

## Financial Benefits and Economic Opportunities

The long-term financial benefits of green energy investments have been highlighted, including reduced energy costs, increased property values, and favorable returns on investment. By capitalizing on federal tax credits, state incentives, and utility-sponsored programs, individuals and businesses can achieve economic savings while contributing to a more energy-independent future.

## Building Resilience and Innovation

Energy storage solutions, such as batteries and smart home technologies, have emerged as critical components in enhancing energy resilience and grid stability. These innovations empower consumers to manage energy consumption efficiently, store surplus energy, and contribute excess electricity back to the grid during peak demand periods.

## Looking Ahead

As we conclude this guide, it is clear that the transition to green energy is not just a trend but a necessity. It requires continued commitment from policymakers, businesses, and individuals to accelerate the adoption of sustainable practices and technologies. By embracing green energy solutions, we not only safeguard our planet for future generations but also create opportunities for economic growth, job creation, and technological advancement.

## Take Action Today

Whether you are a homeowner considering solar panels or a business exploring energy efficiency upgrades, your actions today can shape a more sustainable tomorrow. This guide has provided the tools and insights needed to embark on your green energy journey confidently. Together, we can build a world where clean, renewable energy powers our homes, businesses, and communities sustainably.

Let us move forward with determination, innovation, and a shared commitment to harnessing the power of green energy for a brighter, greener future.